RAINFORD C. OF E.
PRIMARY SCHOOL
TEL: 074 488 3281

Science
AND
Technology

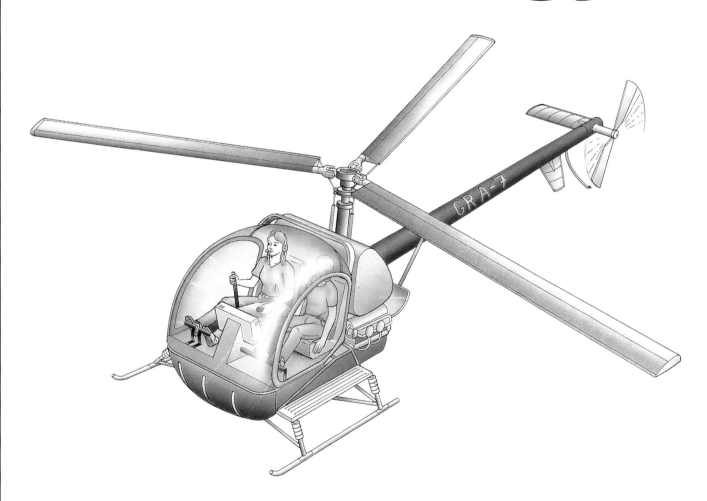

Oxford University Press

Oxford University Press, Walton Street, Oxford *OX2 6DP*

Oxford   New York   Toronto
Delhi   Bombay   Calcutta   Madras   Karachi
Kuala Lumpur   Singapore   Hong Kong   Tokyo
Nairobi   Dar es Salaam   Cape Town
Melbourne   Auckland   Madrid

and associated companies in
Berlin   Ibadan

Oxford is a trade mark of Oxford University Press

© Oxford University Press 1993

All rights reserved. No part of this publication may be reproduced, stored in a retrieval system, or transmitted, in any form or by any means, without the prior permission in writing of Oxford University Press. Within the UK, exceptions are allowed in respect of any fair dealing for the purpose of research or private study, or criticism or review, as permitted under the Copyright, Designs and Patents Act, 1988, or in the case of reprographic reproduction in accordance with the terms of the licences issued by the Copyright Licensing Agency. Enquiries concerning reproduction outside these terms and in other countries should be sent to the Rights Department, Oxford University Press, at the address above.

This book is sold subject to the condition that it shall not, by way of trade or otherwise, be lent, re-sold, hired out or otherwise circulated without the publisher's prior consent in any form of binding or cover other than that in which it is published and without a similar condition including this condition being imposed on the subsequent purchaser.

British Library Cataloguing in Publication Data
Data available

ISBN 0-19-910143-4

Designed by Richard Morris, Stonesfield Design

**Printed and bound in Great Britain by
Butler & Tanner Ltd, Frome and London**

Foreword

Open *Science and Technology* and begin to explore this fascinating world of discovery. Science plays a part in every aspect of our lives, and many of its main themes are touched on here. But you will not find Biology or Earth Science. These are covered in *The Living World* and *Planet Earth*, also in the series.

High-quality text and artwork from the *Oxford Children's Encyclopedia* have been adapted, and expanded with much new material, to produce this easy-to-use reference guide to *Science and Technology*.

How to use *Science and Technology*

Like all reference books you can use *Science and Technology* in two different ways. Make time to sit and browse through it for pleasure, and you will soon find yourself engrossed in a subject you would never have thought to look up.

Science and Technology is organized by subject so, for example, you will find all the Astronomy articles together, and all the Information Technology articles are together, too.

On another occasion you will want to find out about a particular subject, with no time for browsing. Using the index is the quickest way to find out where something is.

```
            R
                                          ———— text information
   radar |139|, |139|————————— illustration
   radiation 15-17,41
      background 71
```

Say you wanted to find out about radar, you must turn to the index, which is always at the back of a book and organized alphabetically. Under the entry **radar** you will find the same page number twice. The first number 139 tells you that you will be able to read about radar by turning to page 139. The second number *139* in *italics*, tells you that you will also find a picture about radar on that page.

Contents

Foreword

Science & materials

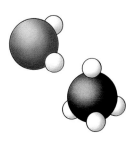

Matter 7
Atoms 8
Elements 10
Chemicals 11
Air 12
Water 13
Carbon 14
Radiation 15
Light 18
Lenses 20
Colour 22
Sound 24
Heat 25
Gravity and forces 26
Machines 28
Time 30
Metals 32
Plastics 34

Space

Space exploration 36
Astronauts 40
Rockets 42
The Moon 44
Solar system 45
Planets 46
Sun 50
Stars 52
Comets and meteors 55
Pulsars, nebulas, quasars 56
Galaxies 57
Constellations 58
Universe 59
Black holes, big bang 60
Observatories 61

Energy and the home

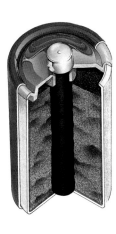

Energy 62
Wind power 65
Solar power 66
Geothermal power 67
Water power 68
Nuclear power 70
Electricity 72
Electricity supply 74
Batteries 76
Magnets and dynamos 77
Household equipment 78
Cookers 80
Cooking 81
Central heating 82
Insulation 83

Transport technology

Transport *84*
Flight *86*
Aircraft *88*
Balloons and airships *90*
Hovercraft *92*
Helicopters *93*
Jet engines *94*
Steam engines *95*
Trains *96*
Railways *98*
Cars *100*
Motor bikes *102*
Internal combustion engines *103*

Trucks *104*
Farm machinery *106*
Bicycles *107*
Sailing ships *108*
Ships *110*
Submarines *113*
Docks and ports *114*
Canals *116*
Bridges *118*
Roads *120*
Structures *122*
Cranes *124*

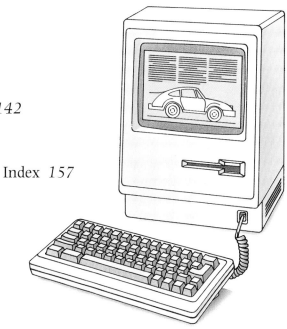

Information technology

Printing *125*
Paper *128*
Cameras *129*
Photography *130*
Radio *132*
Television *134*
Telephones *136*
Satellites *138*
Radar *139*
Calculators *140*
Information technology *142*
Computers *144*

Robots *146*
Lasers *148*
Holograms *149*
Electronics *150*
Recording systems *152*

Index *157*

Acknowledgements

Illustrations & diagrams

Nick Hawken 76t
Oxford Illustrators 7, 8, 9, 11, 12, 14, 15, 16, 17, 20, 21, 22, 26r, 27, 28, 29, 31, 33, 36-37, 38, 41, 42, 44, 45, 51, 54, 55, 59, 64, 66, 67, 69, 70-71, 72, 73, 74-75, 76, 77, 78, 79, 82, 83, 86, 87, 88-89, 90-91, 92, 93, 94, 95, 96, 98, 101, 103, 104, 106, 107, 109b, 112, 113, 117, 123, 124, 127, 129, 132-133, 134, 136, 139, 140, 147, 148, 149, 151, 153, 154, 155
Peter Joyce 24, 25, 26l
Linden Artists: Tony Gibbons 108, 109t, 110-111
Richard Morris 13
David Murray 35, 84-85, 114, 128
Jim Robbins 19, 118-119
Howard Twiner 62-63, 144
Michael Woods 81

Photographs

Abbreviations: t = top; b = bottom; l = left; r = right; c = centre

American Meteorite Laboratory: 57b
Andromeda, Oxford: 46br (NASA), 47c (NASA)
Anglo-Australia Telescope Board: 53, 61t & b
Barnaby's Picture Library: 18r (Adrian Muttitt)
Paul Brierly: 32
Courtesy of British Aerospace: 89
British Museum (Natural History): 11b
Courtesy of British Nuclear Fuels: 17, 71
John Cleare/Mountain Camera: 12
Collections/Brian Shuel: 137t
Courtesy of Diamond International: 14
Royal Observatory, Edinburgh: 57t
Dr Malin, Royal Observatory Edinburgh and Anglo-Australian Telescope Board: 52, 54
Frank Lane Picture Agency: 11t (R.P. Lauwrence)
© Eric Lessing/Magnum Photos: 33
Robert Harding Picture Library: 68, 98t, 98-99b, 99t, 116b
Michael Holford: 30t & b, 31t, 69 (Gerry Clyde), 80b
Hulton Deutsch Collection: 43, 95, 133, 135b, 137b
The Hutchison Library: 141 (Sarah Errington)
Courtesy of the Japan Ship Centre: 108
Kobal Collection: 146tl
Courtesy of Leverton Caterpillar: 105t & b
Mary Evans Picture Library: 97
Millbrook House/Ron Ziel: 96
Richard Morris: 131 sequence
Museo Nazionale Naples/Photo Alinari: 126b
National Maritime Museum, London: 112
National Motor Museum, Beaulieu: 100t, c & br, 102l & br
National Optical Astronomy Observatories: 51, 58b
NASA: 47t, 50-51
Courtesy NCR: 143t
OUP (Chris Honeywell): 28tl, cr & b, 34, 126t, 135t
Courtesy of Philips: 80t, 142l
Planet Earth Pictures: 113 (Peter Scoones)
Popperfoto: 41b
Quadrant Picture Library: 88t, 150
Courtesy of Raleigh: 107
© Redferns: 152 (David Redfern)
Rex Features/François Duhamel/Mega: 64
Peter Roberts Collection/Austin Rover: 100bl
Courtesy of Pickfords Removals: 104b
Science Museum, London: 79b, 88b
Science Photo Library/NASA: 27, 40, 41b, 44, 49, 58t
Science Photo Library: 11c, 15 (Science Source), 56 (ROE & AATB), 60 (Julian Baum), 65 (Tim Davis), 67 (Angela Murphy), 72 (Gordon Garradd), 119b (Harvey Pinics), 138 (Jerrican), 140 (Martin Dohrn), 142-143 (Paul Shambroom), 144 (Martin Dohrn), 145 (Dr Jeremy Burgess), 146br (Tom McHugh), 147 (Peter Menzel), 149 (Philippe Plailly), 151 (Don Thomson)
Spectrum Colour Library: 122 (Dr J Heaton)
Courtesy of Sony Broadcast & Communications: 134
Courtesy of the Sutcliffe Gallery: 130t
Courtesy of Suzuki: 102t
Courtesy of Swatch: 31b
US Geological Survey, Flagstaff, Arizona: 47b
By Permission of the Keeper of the National Railway Museum, York: 99b
Zefa: 18l, 46l, 87, 104t, 115, 116t, 118, 119t, 120l, 120-121, 121, 130b.

Matter

Matter is the scientists' word for what everything is made of. There are three types (called states) of matter: solid, liquid and gas. You are a mixture of all three. You have solid bones and teeth, liquid as the main part of your blood and gas (air) in your lungs.

Solids have a fixed shape, like a cup or a brick. They keep their shape when you pick them up, though some will break if you drop them. Liquids have no fixed shape. They will take on the shape of the container you pour them into. If a liquid is not in a container it flows all over the place. Gases spread out even more easily; a gas leak at a cooker in the kitchen can soon be smelled all over the house. Gases can also be squashed into a small space. Think of pumping air into a bicycle tyre.

Changing state

Most things around us are found only in one state: wood is solid, vinegar is liquid and air is gas. It is only water that we can easily find in the three states: solid, as ice; liquid, as from a tap; and gas, as steam from a kettle. When the solid melts it becomes a liquid. When the liquid evaporates it becomes a gas. As the temperature falls, the gas condenses to form the liquid, then the liquid freezes to form the solid.

Like most substances, water is made up of tiny molecules. Each molecule is a group of atoms. Molecules in a solid are packed close together and hardly move, like balls packed tightly in a box. Tiny electrical forces help to keep them together. Heating gives the molecules more energy and makes them move backwards and forwards more quickly, so that they bump into each other and start to push each other apart. This is why most solids expand when they are heated.

With still more heat, the solid melts into a liquid. The molecules are still close together but now they can move about and change places. This is how a liquid changes shape to fit the vessel it is in.

Further heating makes the molecules separate from each other. Then they speed about as a gas and can spread out to fill whatever space they have.

▲ Heat gives the water molecules enough energy to escape from the liquid and form a gas.

► Ice molecules arranged in a fixed crystal pattern that takes up a lot of space.

▲ Water molecules have enough energy to allow the water to spread out to fit the jug.

A 1 mm cube of common salt contains approximately 20,000 million million million atoms of chlorine and the same number of atoms of sodium.

When water turns to steam, its volume increases by 1700 times. This is why steam can lift the lid off a kettle.

PATTERNS OF ATOMS

The way that matter behaves depends on how its atoms are arranged as well as on which atoms are in it. For instance, one of the hardest materials is diamond and one of the softest is graphite (the 'lead' of a pencil). Both consist entirely of carbon atoms, but the atoms are arranged in different ways. Diamond is so hard that it can be cut into very fine shapes, while graphite is so soft that it leaves a trail as it is pressed across the paper. Coal is also made up mostly of carbon which has been put under great pressure deep in the Earth. It is hard, but not as hard as diamond. Diamond can easily scratch the surface of coal.

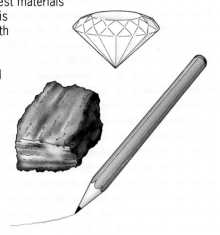

Atoms

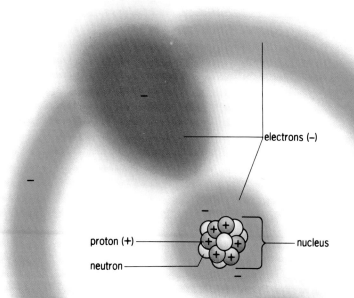

▲ Model of a carbon atom. Electrons behave like smeared-out clouds of electric charge as they move at high speed around a nucleus made up of neutrons and protons. The electrons are all the same, but different colours have been used to show different types of movement.

Protons and neutrons are each 2,000 times heavier than electrons.

Proton numbers:
Hydrogen   1
Carbon   6
Nitrogen   7
Oxygen   8
Iron   26
Gold   79
Lead   82
Uranium   92

All matter is made from about 100 simple substances called elements. Elements are made up of atoms, a single atom being the smallest amount of an element you can have. Atoms are far too small to be seen with any ordinary microscope. More than 4,000,000,000 would fit across the dot on this letter i.

Your body is mainly made from atoms of carbon, oxygen, hydrogen and nitrogen. However, you do not look like any of these elements. That is because atoms can form completely different materials depending on the way they are joined together.

Inside atoms

Atoms are themselves made up of smaller particles called protons, neutrons and electrons. Strong forces bind the protons and neutrons together to form a nucleus at the centre of the atom. The electrons, which are much lighter, move around this nucleus at very high speed. Atoms usually behave like tiny, solid bits of matter, but they are largely empty space. If an atom were the size of a concert hall, its nucleus would be no bigger than a grain of salt. Yet most of the mass is concentrated on the nucleus.

The different types of particle in an atom have different electrical charges. Electrons have a negative (−) charge; protons have a positive (+) charge, while neutrons have no charge. A single atom has the same number of electrons as protons. So, overall, it is uncharged. However, some atoms gain or lose electrons when joined to other atoms. This leaves them either positively or negatively charged. Charged atoms are called ions.

Drawings of atoms often show electrons orbiting a nucleus rather like planets orbiting the Sun. This can be a useful way of thinking about an atom (scientists call it a model) but it is not really a true picture. Electrons behave more like smeared-out clouds of charge, and there is no way of telling exactly where they are at any one instant. A model like this is illustrated on this page. However scientists know that they cannot really describe atoms with pictures. Instead, they have to use the mathematical equations of quantum mechanics.

Different types of atom

All atoms are roughly the same size, but they do not all have the same number of particles. For example, atoms of hydrogen, the lightest element, have just one proton in the nucleus. Atoms of uranium have 92. All atoms of a particular element have the same number of protons (called the proton number or atomic number). However, they may have different numbers of neutrons. These different versions of the same element are known as isotopes. Scientists name isotopes by putting a number after the name of the element, for example: uranium-235. In this case, the total number of neutrons and protons in the nucleus is 235.

Energy from the nucleus

Some types of atom have arrangements of neutrons and protons which are unstable. In time, their nuclei change by shooting out tiny particles and, sometimes, bursts of gamma waves as well. Scientists say that these atoms are radioactive. The particles and gamma waves they then shoot out are called nuclear radiation. Nuclear radiation carries energy which was originally stored in the nucleus. A single atom does not release much energy, but billions of them can heat up materials which absorb their radiation. The high temperatures deep in the Earth are mainly due to radioactivity in the rocks there.

In a nuclear reactor, neutrons are used to 'unlock' the energy stores in nuclei of uranium-235 atoms. The process works like this. A neutron strikes a uranium-235 nucleus and makes it split. As the nucleus splits, it shoots out more neutrons (and some gamma radiation). These neutrons may split other uranium-235 nuclei...and so on in a chain reaction. The splitting process is called fission. It releases energy which makes the materials in the reactor heat up. In a nuclear power station, a carefully controlled chain reaction gives a steady supply of heat. In nuclear weapons, an uncontrolled chain reaction gives an almost instant release of energy. During fission, each split nucleus becomes the nuclei of two new atoms.

▶ Nuclear fission: the two parts of the nucleus, the gamma radiation and the neutrons all shoot off carrying enormous amounts of energy.

MOLECULES

Everything is made of tiny atoms, which are usually linked together in groups called molecules. Some molecules contain the same kind of atoms. A molecule of oxygen gas contains two oxygen atoms. Other molecules contain a mixture of different atoms. The materials made of these molecules are quite different from those made of the separate atoms. Molecules of water contain an atom of oxygen gas and two atoms of hydrogen (another gas). Liquid water and ice are not at all like either of these!

▼ Scientists use models like these to show how the atoms group together to form molecules.

○ hydrogen atom
● oxygen atom
● carbon atom

molecule of oxygen gas

molecule of water

molecule of methane gas

Holding molecules together

In solid materials, the molecules are held together very tightly so the material keeps its hard shape. In liquids like water the molecules are held much more loosely. They do not have a fixed shape but flow to the bottom of their container. Molecules in a gas are free to speed about all over the place, spreading out to fill the space they are in.

Giant molecules

The simplest molecules contain just two or three atoms. However, there are much larger molecules made of many different atoms joined together in complicated shapes. Plants and animals have giant molecules that are made of smaller molecules joined together. These can contain thousands or even millions of molecules, linked into long chains or sheets or spiral shapes.

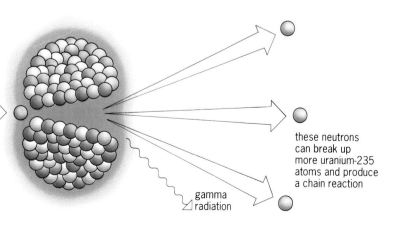

neutron splits uranium-235 nucleus

gamma radiation

these neutrons can break up more uranium-235 atoms and produce a chain reaction

Elements

▼ There are over 100 known elements. Each has its own chemical symbol. About 90 elements exist in nature. Some elements (those in *italics*) have been made in laboratories.

The elements are the basic materials that everything on the Earth and in the whole Universe is made of, including us. Everything is made of tiny atoms, much too small to see, and though many materials contain different kinds of atoms, each element contains only one kind of atom. So an element cannot be split up into different materials.

When discovered	The elements: names and chemical symbols					
Prehistoric times	Carbon	C	Iron	Fe	Silver	Ag
	Copper	Cu	Lead	Pb	Sulphur	S
	Gold	Au	Mercury	Hg	Tin	Sn
Before AD 1650	Antimony	Sb	Arsenic	As		
1650–1699	Phosphorus	P				
1700–1749	Cobalt	Co	Platinum	Pt	Zinc	Zn
1750–1799	Bismuth	Bi	Nickel	Ni	Titanium	Ti
	Chlorine	Cl	Nitrogen	N	Tungsten	W
	Chromium	Cr	Oxygen	O	Uranium	U
	Hydrogen	H	Strontium	Sr	Yttrium	Y
	Manganese	Mn	Tellurium	Te	Zirconium	Zr
	Molybdenum	Mo				
1800–1849	Aluminium	Al	Iodine	I	Rhodium	Rh
	Barium	Ba	Iridium	Ir	Ruthenium	Ru
	Beryllium	Be	Lanthanum	La	Selenium	Se
	Boron	B	Lithium	Li	Silicon	Si
	Bromine	Br	Magnesium	Mg	Sodium	Na
	Cadmium	Cd	Niobium	Nb	Tantalum	Ta
	Calcium	Ca	Osmium	Os	Terbium	Tb
	Cerium	Ce	Palladium	Pd	Thorium	Th
	Erbium	Er	Potassium	K	Vanadium	V
1850–1899	Actinium	Ac	Helium	He	Radium	Ra
	Argon	Ar	Holmium	Ho	Rubidium	Rb
	Caesium	Cs	Indium	In	Samarium	Sm
	Dysprosium	Dy	Krypton	Kr	Scandium	Sc
	Fluorine	F	Neodymium	Nd	Thallium	Tl
	Gadolinium	Gd	Neon	Ne	Thulium	Tm
	Gallium	Ga	Polonium	Po	Xenon	Xe
	Germanium	Ge	Praseodymium	Pr	Ytterbium	Yb
1900–1949	Americium	Am	Francium	Fr	Promethium	Pm
	Astatine	At	Hafnium	Hf	Protactinium	Pa
	Berkelium	*Bk*	Lutetium	Lu	Radon	Rn
	Curium	*Cm*	*Neptunium*	*Np*	Rhenium	Re
	Europium	Eu	*Plutonium*	*Pu*	*Technetium*	*Tc*
1950–1954	*Californium*	*Cf*	*Einsteinium*	*Es*	*Fermium*	*Fm*
1955–1960	*Mendelevium*	*Md*	*Nobelium*	*No*		
After 1960	*Lawrencium*	*Lr*	*Rutherfordium*	*Rf*		

Elements and compounds

You will not have heard of many of the elements in the chart here. In fact most familiar everyday materials are not pure elements but combinations of elements, called compounds. To make a compound the atoms of two or more elements join together to make a completely different material. Water is a compound. It is made from atoms of hydrogen and oxygen. Hydrogen and oxygen are both gases and water is nothing like either. Sugar is another compound. It is made from atoms of hydrogen, oxygen and carbon.

Scientific shorthand

Scientists have given each element a name and a chemical symbol. You can use the symbol if you do not want to write the name in full. Chemists use symbols to show which elements are present in different compounds. For example, a molecule of water contains two hydrogen atoms and one oxygen atom, so the chemical formula for water is H_2O.

Isotopes of elements

Although the atoms of any element all behave the same way, they are not always identical. At the centre of each atom there is a nucleus containing particles called protons and neutrons. All the atoms of an element have the same number of protons but they may not all have the same number of neutrons. These different versions of the same element are called isotopes.

Solids, liquids and gases

At ordinary temperatures most elements are solids. Only 2 elements are liquids and 11 are gases. Most liquids, such as water and oil, are compounds or mixtures of compounds.

Chemicals

When people talk about chemicals, they usually mean substances used in chemistry. There are thousands of these. Some, such as calcium and carbon, are solids, and some, such as sulphuric acid and ethanol, are liquids. Others, such as hydrogen and oxygen, are gases. The whole world is made of substances on their own or mixed with others. So every substance is really a chemical.

▲ Farmers spray chemicals called pesticides on their crops to kill unwanted insects or other pests.

Uses of chemicals

Chemicals have many uses. Some are used as medicines, others stop food going bad, or add flavour and colour to it. Farmers use chemicals to kill pests and weeds. Fertilizers feed the soil and help plants to grow better. Other chemicals are used to get metals out of rock or to make plastics.

Sources of chemicals

The chemical industry gets its chemicals from many sources, including the sea, plants and rocks. Many of our most important chemicals are produced in factories from materials found in oil and coal.

Acids and alkalis

An acid is a sour-tasting substance. Lemons taste sour because they contain citric acid. Vinegar contains ethanoic acid. These are weak acids. Strong acids are far too dangerous to taste or touch. They are corrosive, which means that they can eat into skin, wood, cloth and other materials. Acids are used in many industrial processes, including making explosives, fertilizers and car batteries.

Alkalis are chemicals that feel soapy. They can be strong enough to burn your skin, much like strong acids. Alkalis are used to make soap and glass, and in the chemical industry.

Dangerous chemicals

 radioactive

 dangerous

 flammable

EXAMPLES OF COMMON CHEMICALS

Common name	Scientific name	Chemical formula
Salt	Sodium chloride	$NaCl$
Baking soda	Sodium bicarbonate	$NaHCO_3$
Caustic soda	Sodium hydroxide	$NaOH$
Lime	Calcium oxide	CaO
Chalk	Calcium carbonate	$CaCO_3$
Plaster of Paris	Calcium sulphate	$CaSO_4$

CRYSTALS AND MINERALS

◀ In crystals of salt the atoms are arranged as if they are at the corners of a cube.

The atoms that make up a solid substance are arranged in a special pattern. Some solids have special geometric shapes. We call them crystals.

Minerals are the non-living building materials of rocks. Every mineral has its own chemical structure and crystal shape. Over 2,500 different minerals have been discovered. They include such everyday substances as asbestos, rock salt, the graphite used as pencil lead, the talc in talcum powder and the china clay used to make crockery. Gold, silver, diamonds and other gemstones, and the ores of metals such as copper, tin, iron and lead are also minerals.

▶ Sulphur

Air

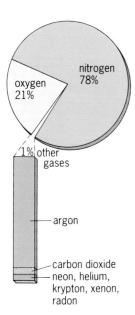

▲ Gases in the air.

You are surrounded by a mixture of gases you breathe called air. You cannot see, smell or taste air, but you can feel it moving when the wind blows. Without air our planet would be a waterless, empty desert without any living creatures.

Air mainly consists of two gases: oxygen (21 per cent) and nitrogen (78 per cent). There is a small amount (less than 1 per cent) of argon and an even smaller amount of carbon dioxide. Carbon dioxide is very important as it is the main food of green plants. The Sun shines on the plants and by a process called photosynthesis helps the plants to use the carbon dioxide to produce oxygen. Oxygen in turn is needed by animals and humans. They breathe in oxygen and use it to provide them with energy. There are all sorts of other things in the air: dust, water vapour, pollen, seeds, tiny microscopic animals, bacteria and pollution from factory chimneys.

▲ Mountaineer wearing oxygen mask.

In 1777 the French scientist Lavoisier showed that air was a mixture of gases and about 1/5 of it was oxygen.

Air can be changed into a colourless liquid under pressure, at about −200°C.

A method for liquefying air on a large scale for commercial use was developed in 1895 by two scientists working separately: William Hampson, an Englishman, and Karl von Linde, a German.

In 1892 James Dewar invented special flasks for maintaining the liquefied gases of air at low temperatures. We call them thermos flasks and use them for keeping things hot.

Mountain air

On mountains the air is much cleaner with less dust and pollution. But as you climb up a mountain the air becomes thinner, with less oxygen. Above 3,000 m (10,000 ft) you have to walk slowly until your body gets used to working with less oxygen. On very high mountains there is so little oxygen that climbers must carry their own supply in cylinders.

Weight and pressure

A bucketful of air weighs about the same as two pages of this book. This may not sound very much. But it means that in a space as big as your school hall there is probably more than a tonne of air. That is more than the weight of a small car!

The weight of the air above is always pressing on us. At sea-level, air pressure is equivalent to the weight of about 1 kilogram pressing on every square centimetre (15 lb on every square inch). However, we are not squashed by this pressure because we have air inside us as well as outside.

Air pressure can be increased by pumping more and more air into the space available. Tyres are filled with extra air so that they can support bicycles, cars and even aircraft. Air under pressure operates fast-acting brakes on trains and lorries. Pneumatic drills are driven by compressed air. Jet engines also rely on air. They get their thrust by sucking in a huge mass of air and pushing it out behind at high speed.

Barometers

Air pressure can be measured with a barometer. The reading is often given in millibars (mb) or in kilopascals (kPa). At sea-level, air pressure is just over 1,000 mb (100 kPa) on average, but this varies slightly from day to day depending on the temperature and the amount of moisture in the atmosphere. Air pressure also changes if you go up in an aeroplane.

Barometers are used to measure the height of an aeroplane and to help in weather forecasting. A fall in pressure warns you that rain may be on the way. Higher pressure usually means fine weather.

Water

Almost three-quarters of the Earth's surface is covered with water, and in places the oceans are miles deep. Frozen water forms the ice-caps at the North and South Poles and snow covers the highest mountain ranges the whole year round. In the sky, massive clouds of water vapour bring rain to us; and where the rain falls and the rivers flow, plants and animals thrive. Without water there would be no life on our planet.

H_2O

Like many other substances, water is made up of molecules. These are so small that even the smallest raindrop contains billions. Every molecule of water consists of two atoms of hydrogen joined to a single atom of oxygen. Scientists say that the chemical formula for water is H_2O.

Water and you

Two-thirds of your body is water. Most of your blood is water. Every organ of your body — brain, heart, liver, muscles — contains water. Press your skin and feel how bouncy it is compared to pressing a sheet of paper. Skin is like a layer of tiny water-filled balloons.

Every day your body loses lots of water. About a litre goes down the toilet. Another half a litre disappears as sweat and in the air you breathe out. On a cold day you can see this water vapour in your warm breath. We must replace the water we lose. We need a litre and a half (2½ pints) of water every day to stay alive. Much of that water comes from our food. Many vegetables and fruit are three-quarters water, and even a slice of bread is one-third water.

▶ A family of four uses about 3,500 litres of water every week. The bar graph shows how much water can be used each time in washing and other activities.

Plants and water

Plants need water to grow. They use water and other chemicals to make the substances needed for new plant growth. They use water to carry substances between their roots and leaves. And they use the pressure of water in their cells to stay firm and rigid. Without water, plants wilt. Plants usually take in water through their roots and lose it through pores in their leaves.

Different kinds of water

Drinking water can taste completely different depending on where you live. The water that comes out of the tap comes from rainfall that washes over rocks and flows along rivers. On its travels, the rain-water dissolves gases from the air and many different substances from the rocks. So the taste of our drinking water depends very much on what is in the local rocks.

Ice

Water that is frozen solid is called ice. Water is unusual — when it freezes, it expands and takes up more space. A bucketful of ice therefore weighs less than a bucketful of water. When a pond freezes, at 0°C, the ice floats at the top. Water is actually heaviest at 4°C (39°F).

Water expands when it freezes, and can break pipes.

Ice will float on water. When a pond freezes in winter, fish can survive in the water below.

The freezing point of water was taken as the bottom of the Celsius (centigrade) scale of temperature, so that water freezes at 0°C. The boiling point of water was taken as 100°C.

toilet flushing (10 litres) — shower (30 litres) — bath (80 litres) — washing machine (100 litres) — dishwasher (50 litres)

Carbon

Carbon is the black material you see on burnt wood or toast. Carbon can also form two very different types of crystal. One is graphite, the black material used as the 'lead' in pencils. The other is diamond, the hardest substance known.

When carbon combines with hydrogen, it forms a whole family of new materials called hydrocarbons. These include fuels such as natural gas, petrol, paraffin and diesel oil. Carbon also combines with hydrogen and oxygen to form foods like sugar and starch. These are called carbohydrates. When foods and fuels are burned, the carbon in them combines with oxygen to form carbon dioxide gas. Vinegar, alcohol, perfumes, plastics and disinfectants also contain carbon.

Living carbon

Plants and animals are mainly made from materials containing carbon. So carbon is essential for life. The air around us contains tiny amounts of carbon dioxide gas. Plants take in carbon dioxide through their leaves and water through their roots. Using the Sun's energy, they turn the gas and water into new plant material. So plants are partly carbon. Animals like us eat plants for food, so our bodies are also partly carbon. When we 'burn up' our food, carbon is used in making carbon dioxide gas.

Fuels like petrol also contain carbon. They were formed from the decayed remains of plants and animals which lived millions of years ago.

▲ A diamond cut into the shape of a pear.

Grass is about 4% carbon (by weight).

Human beings are about 20% carbon.

Coal can be over 90% carbon, depending on the type.

Charcoal is almost 100% carbon.

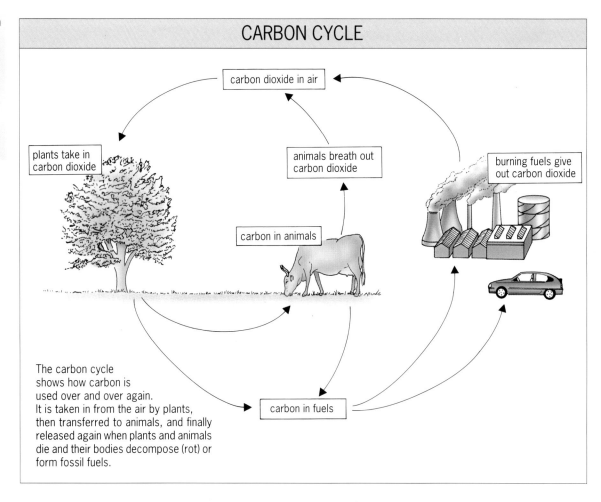

CARBON CYCLE

The carbon cycle shows how carbon is used over and over again. It is taken in from the air by plants, then transferred to animals, and finally released again when plants and animals die and their bodies decompose (rot) or form fossil fuels.

Radiation

Radiation is energy on the move. Some kinds of energy, such as heat and light, travel about as invisible waves. Other kinds of radiation are tiny particles that shoot out from atoms at enormous speeds. Cosmic rays from space are particles. Radioactive materials can produce a mixture of radiations, some of which are particles and some waves.

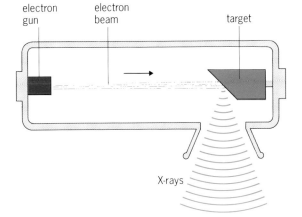

◀ Electric current is a stream of tiny particles called electrons. In an X-ray tube, X-rays are given off when a beam of electrons hits a metal target.

Radioactivity

When people talk about the radiation from nuclear power-plants, they mean the mixture of particle and wave radiation that comes from radioactive materials. These materials are made up of atoms which change into a different kind of atom by throwing out tiny atomic particles.

There are two kinds of particles, called alpha and beta. Each beta-particle is an electron. Each alpha-particle is two protons and two neutrons. Radioactive materials also throw out gamma rays. These travel as waves, carrying a lot of energy. Scientists often use a Geiger counter to measure the nuclear radiation from radioactive materials.

Useful radioactivity

Nuclear radiation has many uses. Doctors use it to kill the cells that make cancer growths. Radioactive materials can help doctors to find out how well the kidneys and other parts of the body are working. They have to be very careful not to use too much nuclear radiation or it would harm the patient. Space probes that travel far from the Sun use radioactive materials to make electricity. Radioactive materials are also used as portable sources of the powerful gamma rays that can be used to X-ray steel girders in buildings or bridges to locate cracks.

X-rays

X-rays are a kind of high energy radiation that can penetrate solid objects. When X-rays hit a photographic film or plate, they make it go dark. So X-rays can be used to take photographs of the inside of solid objects. They cannot penetrate very hard substances, so these appear pale on the photograph. Doctors and dentists use X-rays to take pictures of broken bones and growing teeth inside your body.

Radioactive materials
About 50 are found naturally. Uranium is the best-known. Over 2,000 can be made artificially. One of these is plutonium, used to make nuclear bombs.

Speed of light
All electromagnetic waves, including light, move at 300,000 km/sec (186,000 miles/sec). Nothing can go faster than this.

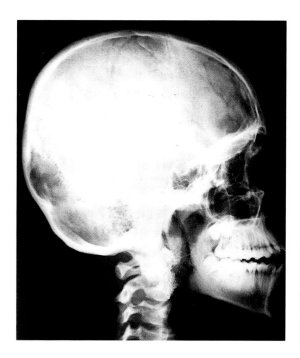

◀ X-ray of a normal human skull. X-rays make shadow pictures because they go through soft parts of your body, like skin, but are stopped by the hard bones or teeth.

Using X-rays

X-ray photographs are used to detect cracks and fractures in bones. But doctors also use much stronger X-rays to treat cancer. These are carefully measured and aimed to kill the cancer without damaging the rest of the body. In industry, X-rays can check inside machinery for cracks or faults. At airports, they are used by security staff to check luggage for weapons and bombs. Scientists use X-rays to study how atoms are arranged inside solid materials like crystals. Astronomers can learn more about the stars by studying the X-rays that come from them.

Wavelength and frequency

Every type of wave radiation has a wavelength and a frequency, just like the ripples you see on water. Imagine you are by a pond. Someone drops a stone in so that tiny waves travel towards you across the water. The number of waves reaching you every second is called the frequency. The distance from one crest to the next is the wavelength. The more waves reach you every second, the closer together are the crests; the higher the frequency, the shorter the wavelength.

Electromagnetic radiation

Most types of wave radiation belong to the same family. They are electromagnetic waves. You can see the different types in the chart below. They can all travel through empty space, which is why light and heat can reach us from the Sun. And they all travel at the same speed, the speed of light. But they all have different wavelengths (and frequencies).

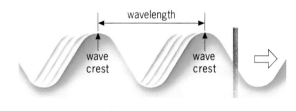

▶ The wavelength is the distance between neighbouring crests.

▶ Diagram showing how the different members of the family of electromagnetic waves are related.

Wavelength and frequency on the radio

Long wave
If frequency is 200 kHz (200,000 waves reach your radio every second) wavelength is 1,500 m.

Medium wave
For frequency of 1,000 kHz (1 million waves reach your radio every second) wavelength is 300 m.

FM (VHF)
For frequency of 100 MHz (100 million waves reach your radio every second) wavelength is 3 m.

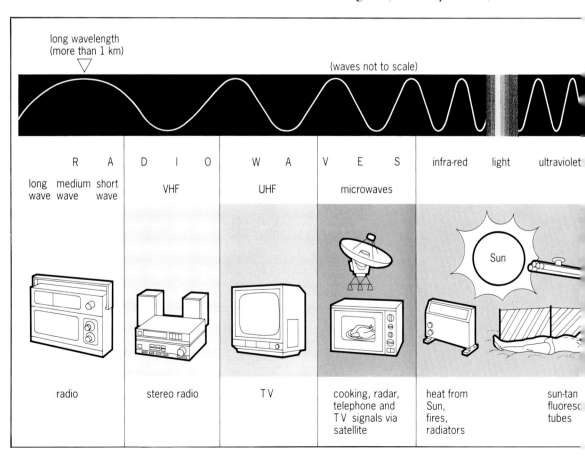

SCIENCE & MATERIALS

Cosmic rays

Out in space, speeding around almost as fast as light travels, are cosmic rays. This kind of radiation is made up of tiny atomic particles. They carry a great deal of energy because they are travelling so fast. Astronomers think that many cosmic rays come from exploding stars called supernovas, though some come from the Sun, and some from other far-away galaxies.

Sound radiation

Sound is a different type of radiation. When you speak, sound waves radiate out from your mouth and are picked up by other people's ears. The sound waves are tiny compressions that travel through the air.

Low notes have the lowest frequencies and the longest wavelengths. High notes have the highest frequencies and the shortest wavelengths. Sounds which are too high for the human ear to hear are called ultrasonic sounds.

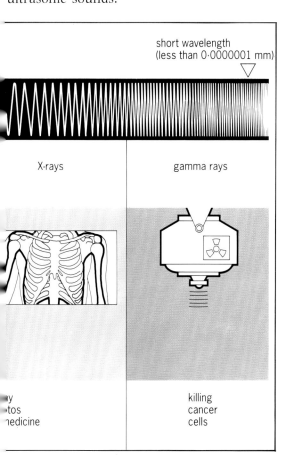

Radiation dangers

There are small amounts of nuclear radiation around us all the time which come from radioactive materials in the Earth. This is called 'background radiation' and normally does us no harm. However, scientists need to keep a close check so that suitable precautions can be taken, as too much radiation can be dangerous. For example a natural radioactive gas called radon leaks up through the ground in some places and can collect in houses. Pumps have to be fitted to get rid of it.

A nuclear bomb or an accident at a nuclear power-plant makes enough radiation to kill people nearby. Radioactive dust can be carried hundreds of miles on the wind. This radiation may cause cancer, often many years later. So nuclear power-plants have huge concrete walls to stop the radiation escaping, and people who work in them have to wear special clothing to protect them.

Protection from radiation

Alpha-particles are stopped by a sheet of paper.

Beta-particles are stopped by 3 mm of metal or 6 mm of wood.

Gamma rays are stopped by 5–10 cm of lead or 30–60 cm of concrete.

Sound Sound waves travel at a speed of about 330 metres per second (740 mph), which is nearly a million times slower than the speed of light.

◀ Worker exposed to radiation wearing protective clothing.

A new science

Radioactivity and the invisible electromagnetic radiations were first discovered by pioneer scientists in the last century.

Heinrich Hertz discovered radio waves in 1887.
Wilhelm Röntgen discovered X-rays in 1895.
Henri Becquerel discovered in 1896 that uranium gave off radiation – it was radioactive.
Marie Curie found the radioactive elements polonium and radium in 1898.

Light

Speed of light
Light travels through space (and air) at a speed of 299,000 km per second (186,000 miles per second). If you could travel at that speed, you could go more than seven times round the world in a second!

Scientists believe that nothing can travel faster than light. That is because there is not enough energy in the whole universe to make even the smallest thing reach that speed.

▼ A prism splits white light into the colours of the spectrum. Most people think they can see six separate colours: red, orange, yellow, green, blue, violet. Really, the spectrum is a steady change of colour from beginning to end.

Light is a kind of radiation, travelling at very high speed. It comes from the Sun or electric lights or from hot or burning things, like a fire or a flame. It travels through transparent things like glass or water, but it bounces off things that look solid, and they reflect the light. We see these things when some of this reflected light enters our eyes. Without light we cannot see anything. Shiny surfaces reflect almost all the light, while darker, rougher ones reflect less and absorb more light. Tiny particles like dust or smoke scatter light in all directions.

Bending light

Light travels more slowly in water or in glass than in air. When a beam of light goes from water into the air, the part in the air will start to go faster than the part still in the water. This can bend the light. It makes a straw leaning in a glass of water look bent, and a swimming pool look less deep than it really is.

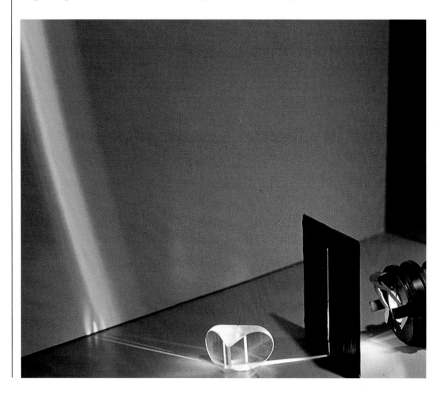

The spectrum

The light from the Sun is really a mixture of colours, and a prism (a pyramid-shaped piece of glass) will split it up into its different colours. The glass bends the light, but some colours bend more than others and they spread out. The range of colours is called a spectrum.

Waves and particles

Light speeds around, bouncing off things as it goes. It travels in straight lines as if it were a stream of tiny particles. Scientists call these particles photons. Shadows show us that they travel in straight lines. If light could curve round things it would not cast shadows. Sometimes light seems to travel like waves on water. If you look at tiny waves, called ripples, travelling across the surface of a pond, you can see that they have peaks at the top and dips at the bottom. The distance between two peaks is called the wavelength. Light waves are electric and magnetic ripples which can travel through space. Their wavelengths are so small that a thousand of them would take up less than a millimetre. The different colours in the spectrum have different wavelengths. Red light waves have the longest and violet the shortest.

▲ Studying ripples on the surface of water can help us understand how light waves behave.

Lighting

Nowadays in Western countries, most lighting is electric. But in the past people used lamps which burnt fat, oil or gas. And these are still used in many parts of the world.

In England, town councils are now responsible for lighting streets. But in the 15th century householders had to hang lanterns outside their homes at night. The candles inside were easily blown out or stolen. So people often carried torches made of flax dipped in pitch to light their way. Oil-lamps were used in London from the 1680s, and gas lamps were first used for street lighting in 1813. The light was very dim and link-boys carried flaming torches for people who could afford to hire them. After 1830 parishes and towns began to provide lighting. Electric street lighting was first used in 1881 in Godalming.

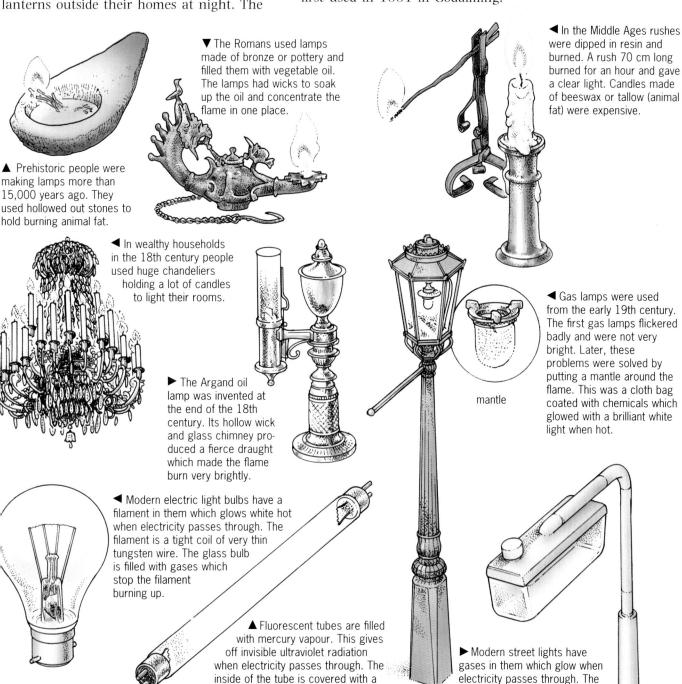

▲ Prehistoric people were making lamps more than 15,000 years ago. They used hollowed out stones to hold burning animal fat.

▼ The Romans used lamps made of bronze or pottery and filled them with vegetable oil. The lamps had wicks to soak up the oil and concentrate the flame in one place.

◀ In the Middle Ages rushes were dipped in resin and burned. A rush 70 cm long burned for an hour and gave a clear light. Candles made of beeswax or tallow (animal fat) were expensive.

◀ In wealthy households in the 18th century people used huge chandeliers holding a lot of candles to light their rooms.

▶ The Argand oil lamp was invented at the end of the 18th century. Its hollow wick and glass chimney produced a fierce draught which made the flame burn very brightly.

◀ Gas lamps were used from the early 19th century. The first gas lamps flickered badly and were not very bright. Later, these problems were solved by putting a mantle around the flame. This was a cloth bag coated with chemicals which glowed with a brilliant white light when hot.

mantle

◀ Modern electric light bulbs have a filament in them which glows white hot when electricity passes through. The filament is a tight coil of very thin tungsten wire. The glass bulb is filled with gases which stop the filament burning up.

▲ Fluorescent tubes are filled with mercury vapour. This gives off invisible ultraviolet radiation when electricity passes through. The inside of the tube is covered with a white powder which glows brightly when the ultraviolet strikes it.

▶ Modern street lights have gases in them which glow when electricity passes through. The colour of the light depends on the type of gas in the tube.

Lenses

Lenses like the ones you see in spectacles are specially shaped pieces of glass or plastic that bend a beam of light as it goes through them. There are two types of lens. A convex lens bulges outwards and is thicker in the middle. A magnifying glass is a convex lens – it makes things look bigger. A concave lens curves inwards and is thinner in the middle. A concave lens makes things look smaller. Spectacles worn by short-sighted people have concave lenses.

Looking through lenses

If you hold a convex lens in front of a sheet of card facing a window, the lens makes a small, upside-down picture of the window on the card. We call this picture an image. The lenses we use most often are the ones in our eyes. They make an image of what we are looking at, inside our eyes. In the same way, a camera lens makes an image on the film inside the camera. Microscopes, binoculars and telescopes use lenses to enable us to see things that are too small or too far away to be seen with our eyes alone.

Telescopes

The simplest telescope is a tube with a lens at each end. This kind is called a refracting telescope. The lens at the front collects the light. The lens you look through is called the eyepiece. It magnifies the image of what you are looking at.

A telescope that uses a curved mirror to collect the light is called a reflector. The main mirror concentrates the light onto another small mirror which then reflects it into an eyepiece. Really big telescopes are always reflectors because it is impossible to support large lenses so that they do not bend under their own weight. The bigger a telescope's collecting mirror or lens, the fainter the things it can see and the stronger

▲ A concave lens spreads rays of light apart. A convex lens brings them together.

▼ A refracting telescope. Rays from the distant planet are caught and focused by the main lens. An image of the planet is formed in front of the eyepiece lens. The eyepiece lens magnifies this image.

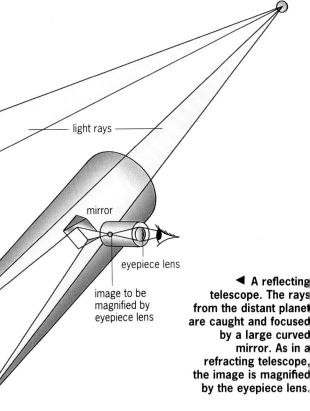

◀ A reflecting telescope. The rays from the distant planet are caught and focused by a large curved mirror. As in a refracting telescope, the image is magnified by the eyepiece lens.

the magnifying power you can use with it. Astronomers like telescopes with mirrors that are as big as possible. The largest reflecting telescope in the world has a mirror 6 metres across.

Radio telescopes are even bigger. They focus radio waves instead of light rays, and can 'see' distant objects in the Universe. Radio waves have much longer wavelengths than light rays, so they need a bigger reflector dish. The world's largest radio telescope 'dish' is 100 m across.

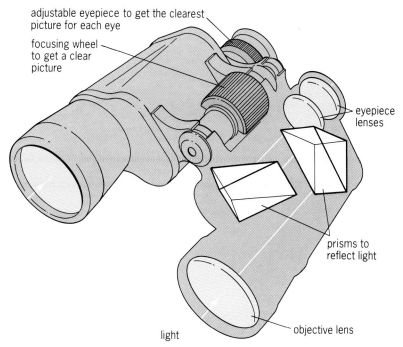

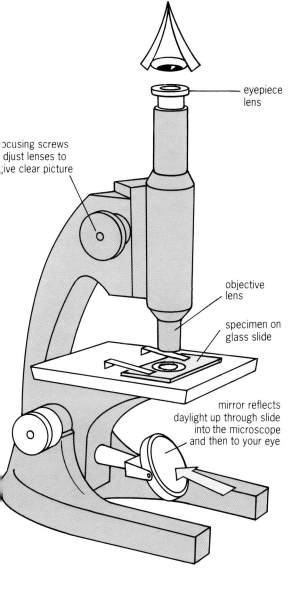

▼ A standard monocular (one eye-piece) microscope.

Binoculars

People often use binoculars when they are bird-watching, to make far-away things, like the birds, seem bigger (magnified). Binoculars are just a pair of telescopes, one for each eye. They have prisms (pyramid-shaped pieces of glass) which reflect the light in a zigzag path. These make the binoculars much shorter than a telescope. The prisms also turn the picture the right way up. With a pair of binoculars you can use both your eyes so that you can judge distances. This is not possible with a telescope.

Microscopes

A microscope makes things look much larger, letting us see things that are too small to see normally. It is more powerful than a magnifying glass because it has at least two magnifying lenses. Even a small microscope can easily magnify things 100 times, so a hair would look 1 cm thick instead of $\frac{1}{10}$ mm. The specimen you want to look at must be very thin because light goes through it up into the microscope. Some things are too small to be seen with an ordinary microscope, but many of these can be seen with the special electron microscope.

▲ Binoculars are really two telescopes side by side, with prisms to keep the length short and turn the image up the right way.

If binoculars are marked 8x30, the magnification is 8; the binoculars make things look 8 times bigger. The diameter of the objective lens is 30 mm (1·2 in). Lenses of bigger diameter collect more light and give a brighter picture.

Opera glasses look like binoculars but they do not need prisms. They use a different lens arrangement which magnifies less and gives a narrower view.

Colour

Light rays themselves are invisible; we can see them only when they bounce off things. In the diagram we see the light bouncing off specks in the glass or smoke particles in the air.

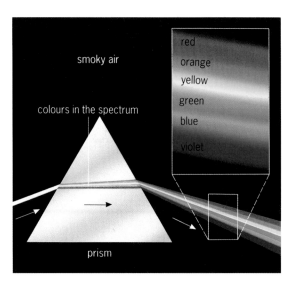

▲ A glass prism splits white light into all the colours of the rainbow. The spread of colours is called a spectrum. It shows up easily in the smoky air.

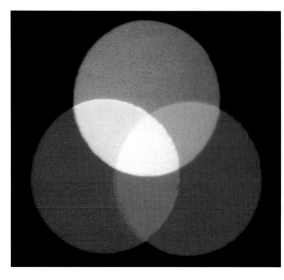

▲ When mixing light beams, the colours are added together. By mixing red, green and blue in varying amounts, other colours can be made.

We live in a world full of colour. Perhaps the most beautiful sight is a rainbow arching its colours across the sky. But how do we see all this colour?

All light travels to us like the waves that ripple across a pond when you drop a stone in. The distance between the top of one ripple and the top of the next is called the wavelength. Light of different colours has different wavelengths. Red light has the longest wavelength and violet the shortest. The wavelength of the light makes us see different colours, but that is not the whole story. The brain has an important part to play in sorting out what we actually see. Nobody knows exactly how we see all the shades of all the different colours.

▼ Mixing of light to give secondary and complementary colours.

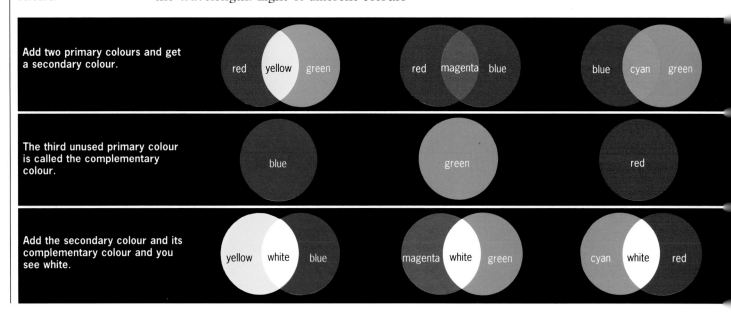

Add two primary colours and get a secondary colour.

The third unused primary colour is called the complementary colour.

Add the secondary colour and its complementary colour and you see white.

White light

The light from the Sun or a lamp bulb seems colourless, but it contains all the colours of the rainbow. Whenever our eyes receive a mixture of all the colours of the rainbow we see this as white light.

Coloured light

To make beams of light of different colours, you can put sheets of transparent coloured plastic (filters) in front of a torch beam of white light. A red filter gives you a red light beam. A green filter gives you a green light beam. If you shine the red beam and the green beam onto a sheet of white paper, you will see yellow where they overlap! When light is added together in this way we need only three basic colours: red, green and blue, in different proportions, to make all other colours. Scientists call these three basic colours the primary colours. Television uses only these colours to produce colour pictures.

Paints and dyes

Perhaps a room has blue walls and a red carpet. These colours come from paints and dyes. But mixing paints (or dyes) gives different colours from mixing beams of light. When you mix beams of light the colours are added together. But, when you mix paints, colours are taken away. White light shines on the paint on the blue walls and the paint absorbs most of the colours. You see only what is left: in this case colours in the blue part of the spectrum. In the same way the carpet looks red because all the other colours in the blue part of the spectrum are absorbed by the dye. If you were to mix red and blue paint, the result would be very dark because, between them, the red and blue paint would absorb almost all the light.

Artists and printers can make most colours by mixing paints or inks of three basic colours: magenta, yellow and cyan (which they sometimes call primary red, primary yellow and primary blue).

◀ Printers use three basic colours and this shows what happens when they are mixed. Yellow takes away the blue light and cyan takes away the red light and so if they are mixed we are left with green. If you mix all three basic colours, then red, blue and green are all taken away and the result is almost black.

Telling colours apart

Not everyone is good at telling colours apart. Some people cannot see the difference between any colours. They are colour blind. Others have difficulty telling red from green. About one boy in ten is born with red–green colour blindness, though hardly any girls are affected. Colour blindness is something which people inherit. When darkness falls, everyone has problems telling colours apart because the bits of the eye which pick up colours work properly only if there is plenty of light.

In everyday life, colours are often used to give warnings or other messages. For example, red may mean 'stop' or 'danger', while green may mean 'go' or 'all clear'. This can cause problems for colour blind people. With traffic lights, they have to look at the position of the light (top or bottom) to know whether it means stop or go. Colour blind people also find it difficult to work with electric cables. In most cables, the wires are in different colours so that electricians can work out which connections to make.

You cannot accurately describe colours but you can match a colour to one on a chromaticity colour chart. These charts have thousands of different colours, and for each they state the exact proportions of the three primary colours needed to make that particular colour.

Colours used to describe people and their moods

red	angry
green	jealous
blue	depressed
yellow	cowardly
white	pure
grey	uninteresting

Sound

Echoes Sound waves can be reflected just like light rays. Reflected sound waves are echoes.

Ultrasound is sound too high-pitched for us to hear. Bats use sounds up to 200,000 Hz to find their way in the dark.

Infrasound is sound too deep for us to hear. Elephants talk to each other over long distances using sounds as low as 1Hz.

A sonic boom is a loud explosive noise heard when an aircraft flies overhead at a speed greater than the speed of sound. It is caused by the shock waves set off as the plane squashes the air in front of it.

Breaking the sound barrier This happens when a plane starts to travel faster than sound.

Everything that can be heard is a sound. The sounds that we hear start when something makes the air vibrate (wobble) backwards and forwards quite quickly. Twang a rubber band and you can see it vibrating and hear a sound. Place a finger on the band to stop the vibrations and there is nothing to hear. When the rubber band is plucked, its vibrations make the air next to it vibrate. Then the air next to that is forced backwards and forwards and so the sound moves outwards away from the band. When the vibrating air reaches your ear, it makes the ear-drum move in and out, and you hear a sound.

Making sounds

Anything that vibrates makes a sound. A bee's wing moves backwards and forwards very quickly and we hear a buzz. When you speak your voice box vibrates the air coming out of your mouth.

How sound travels

All that we hear has travelled through the air around us. Remove the air from the room you are in and you would hear nothing. There is no sound out in space where there is no air.

But sound does not travel only through air. Vibrations can travel through water, glass, brick, concrete and other substances. Vibrations move particularly easily and fast through water. Whales and porpoises make sounds which travel over hundreds of kilometres of oceans.

Typical sound levels in decibels

Loudness of sound

The loudness of a sound is measured in decibels (dB). The closer you are to whatever is producing the sound, the louder it is. If you are close to a very loud sound, like an explosion, your hearing can be damaged. But loud sounds, which may not affect your hearing immediately, can produce serious damage if your ear receives them for a long time. Pop groups and their fans often get hearing damage from standing too close to the loudspeakers. Listening with headphones to loud music can also produce deafness.

Speed of sound

In air, sound travels at a speed of about 330 metres every second (740 mph). That is about four times as fast as a racing car but only half the speed of Concorde. Sound travels slightly faster on a hot day than it does on a cold day. Sound travels much faster through solids and through water than it does through air.

High and low sounds

When a sound is made, the number of vibrations every second is called the frequency. Frequency is measured in hertz (Hz): 1 hertz means one vibration every second. The human ear can hear sounds as high-pitched as 20,000 Hz, and as deep as just 20 Hz.

Heat

Heat is a type of energy. You can see heat in action if you watch water being heated. The still water in the pan becomes a furious bubbling liquid with steam rushing up and out of the pan. Keep clear of the steam because it can destroy your skin and cause burns. If a lid is put on the pan it will jump up and down as the steam tries to push it off.

Heat in action

To understand what is happening when you heat something we need to imagine what it is like inside that substance. Everything is made of very small particles called molecules. In a solid, these molecules are locked very close together but they vibrate to and fro a little. When heated, they vibrate even more and begin to break free from each other. The solid melts into a liquid. With further heating, the molecules move so quickly that they escape from each other. The liquid becomes a gas.

Expansion

Heating a solid makes the molecules vibrate faster. As they vibrate more vigorously they push neighbouring molecules further apart. The result is that the object increases in size. It expands. In bridges, small gaps are left between the long stretches of metal girders to allow space for expansion on a hot day.

Heat and temperature

It is easy to think that heat and temperature are the same but they are not. It takes as much heat to make a litre of water in a kettle boil at 100°C as it does to heat a whole bath of water to a comfortable temperature of about 25°C. The heat is spread out in the larger volume of bath water and so the water is not as hot as in the kettle. If the same amount of heat is concentrated in something very small, such as a tiny piece of metal, then it will make it very hot. It will glow red and have a temperature of more than 700°C.

Radiation

Heat from a fire travels to your skin in the same way as heat and light reach you from the Sun. They are types of radiation and can travel through empty space. The radiation makes the molecules in your skin move faster, and you feel warm.

Conduction

The handle of the spoon sticking out of a cup of hot tea soon feels hot itself. The atoms of the spoon vibrate faster with the energy they get from the hot tea; they bump into their neighbours and make them vibrate faster. In this way heat moves up the spoon. This movement of heat is called conduction. Metal conducts heat easily.

Convection

Hold a piece of tissue over a hot radiator and watch it drift upwards. The air above the radiator is being heated and rises. As the air cools, it becomes heavier and sinks to the floor. This movement of hot air round a room is called convection and is a very important way of heating houses.

Heat and work

Rub two sticks together fast and the heat produced can start a fire. You have to work hard to rub the sticks together because of the friction. The work you put into the rubbing produces the heat. Work produces heat, and of course heat can do work. Steam pressure can drive engines and produce electricity. The study of the way that heat and work can be converted into each other is called thermodynamics.

The kinetic theory of the 19th century proved that heat was produced by the movement of molecules. Before that, heat was thought to be a substance called *caloric* which could be poured or moved from place to place.

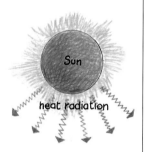

▲ Heat radiation from the Sun travels through empty space.

▲ Heat travels quickly along a metal spoon. Metals are good conductors of heat.

▲ Convection currents quickly carry heat around a room.

Two litres of water at a particular temperature have twice as much heat as a litre of water at the same temperature.

Forces and Gravity

Force is the word we use for a push or a pull. A force will make a still thing move. It will make a moving thing travel faster, more slowly, or in a different direction. Together, two or more forces can make something stretch, squash, bend, twist or turn.

◀ When you ride a bike, some friction is helpful and some is a nuisance.

Friction helps ...

friction helps your hands grip the handlebars

friction stops you when you put the brakes on

friction helps your feet grip the pedals

friction helps the tyres grip the road

Friction is a nuisance ...

air friction slows you down

friction slows the pedals

friction slows the wheels

Friction

Friction is the force which tries to stop one material sliding over another. A moving car will encounter friction with the air and the road. This will slow it down. Parachutists use friction with the air (air resistance) to slow their descent.

The rate at which an object falls through the air depends on its air resistance. A feather falls more slowly than a stone because it has a larger surface area and finds it harder to push the air aside. But a large and a small cannon ball will hit the ground at the same time, despite their different weights.

Tension and compression

If you stretch a rubber band, you feel a force called tension. Tension forces are very useful. They support climbers on ropes and goods in shopping bags. We use tension forces to hold up our clothes! Compression (squashing) forces are the opposite of tension forces. They are produced when you lie on a bed, sit on a chair or just stand on the ground. Without compression forces, nothing could ever support your weight.

Different kinds of force

Scientists think there are 5 kinds of force in the Universe: gravitational, magnetic, electric and two types of nuclear force. They believe that the five types of forces are related, but they still do not fully understand how.

Electric forces hold the atoms together in solids and liquids, so tension, compression and frictional forces are really examples of electric forces.

Gravity

Gravity is the pulling force that holds us all down on the surface of the Earth. The 17th century scientist Sir Isaac Newton worked out the theory of gravitation to explain why apples fall to the ground. Everything has a gravitational pull towards everything else. The more massive the object is, the larger its pull. The Earth, which is huge, has a very strong pull.

▼ Some of the things you use forces for.

lifting

turning

stretching

squeezing

bending

▲ Weight is measured by comparing your mass with the mass of a weight on a scale. In space, the pull of gravity is less, but it would still take the same number of kilograms to balance you on the scales.

SCIENCE & MATERIALS

◀ Astronauts have to cope with weightlessness. Here, astronaut Robert Cabana, pilot of the 1990 Shuttle Mission STS-41, is using a cine camera inside the shuttle.

Kilograms and pounds really measure mass, not weight.

The Earth bulges slightly at the Equator, so you would weigh less there than at the poles.

Gravity, weight and mass

Gravity pulling down on us gives us our weight: it makes us feel heavy. The Moon is smaller than the Earth, so it has a weaker gravity pull. We would weigh 6 times less on the Moon, even though we are still exactly the same size. Jumping would be very easy. Way out in space, far from any large solid objects, our weight would be so tiny that we would float.

The mass of something tells you how much matter there is in it. The pull of gravity (weight) depends on the mass of the object being pulled. So you can find out the mass of something by weighing it. Scientists measure all forces in newtons. On Earth, every kilogram of mass weighs about 10 newtons. An apple weighs about 1 newton, and an average 10-year-old weighs about 300 newtons.

Gravity and the Universe

Without the pull of the Earth's gravity, the atmosphere and the oceans would have floated away into outer space, water would not run downhill and rain would not fall to the ground. The pull of the Earth's gravity holds the Moon in its orbit circling round us, and the Sun's gravity keeps the Earth and the other planets in their orbits. In fact, the Sun and the planets would not exist at all if gravity had not pulled together particles of gas and dust to make them.

Centre of gravity

Gravity pulls strongest at the centre of a object: its centre of gravity. When you are standing up straight, your centre of gravity is directly above your feet. If you lean too far forwards, backwards or sideways, your centre of gravity is no longer above your feet, and you will fall over.

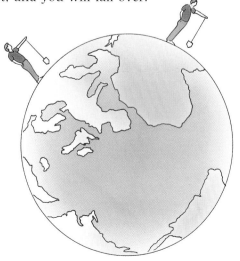

◀ Plumblines all point towards the centre of the Earth because the Earth's gravity is greatest at its centre.

Machines

Machines are devices that make work easier. Your home and school may have machines in them, for washing clothes, preparing food, or moving things around. Some may be electronic: a television or a computer. But there are many machines so simple that we do not even notice them. We do not even call them machines. A door handle operating a latch is one. The hinged door itself is another. Without the hinge it would be hard work moving a large piece of wood out of the way every time you wanted to go in or out of a room.

However big and complicated a mechanical thing may look, most have in them the five simple machines: the **lever**, the **wheel** and **axle**, the **ramp**, the **screw** and the **pulley**. Machines make it possible to put in a force at one place, and get it out at another. They may change the direction of this

Levers can help us lift heavy things. A spade is a lever. A wheelbarrow is another. With a long lever, resting on a pivot, you could lift a heavy weight.

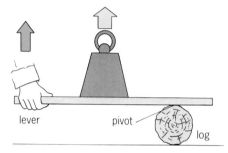

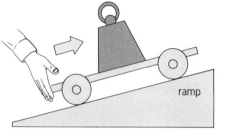

Ramps, or slopes, make it easy to go up a short distance by moving forward quite a long way.

A **screw** is really a long ramp, wound up. Some car jacks use a screw. By turning it round, the weight of the car may be lifted higher.

The **wheel and axle** make it possible to move loads along. Disabled people can get about in wheelchairs. The bigger the wheels, the easier they are to turn. The long spokes act like levers.

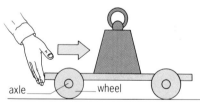

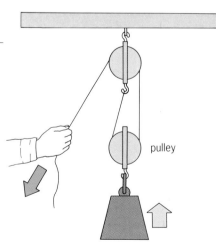

Pulleys raise heavy things by making it possible to pull downwards, which is much easier than pulling upwards. They also magnify the pulling force. Big complicated pulley systems use many sets of wheels. Using a pulley system, one person can lift a car engine.

force. So you may wind the handle of a car jack round and round, but the car goes up. Or they may change the strength of the force. You can move things in a wheelbarrow that you could not possibly carry.

Transmitting power

Simple machines can be joined together to do a complex job. In factories and workshops, machines do work that would take a long time to do by hand. The force to operate the machines is transmitted to them in different ways.

Series of gearwheels reduce the effort needed to do heavy work. Meshing the teeth of two gearwheels (locking them together) makes all kinds of changes possible. Gears can be made to turn faster or slower, and in different directions.

Belts and chains

Belts and bands drive many kitchen machines: the washing machine, the tumble-drier, the food processor. Tiny belts drive a video recorder. Different sizes of pulley turn the machines faster or slower.

If a belt has to transmit too much power, it will slip. A chain overcomes this problem. It cannot slip because it fits over teeth, called sprockets, on the gearwheels. Bicycles use chains.

Control

Machines are becoming more complex and many work too fast for people to control. Electronic devices are now used to control many machines. Electrical sensors can tell when it is time to switch on other parts of the machine. They check that everything is running safely. The machines control themselves: they are robots.

Modern factories may be run almost entirely by robots. They can assemble a car body, welding the pieces together with great accuracy. All that people need to do is to service, and sometimes repair, the robots.

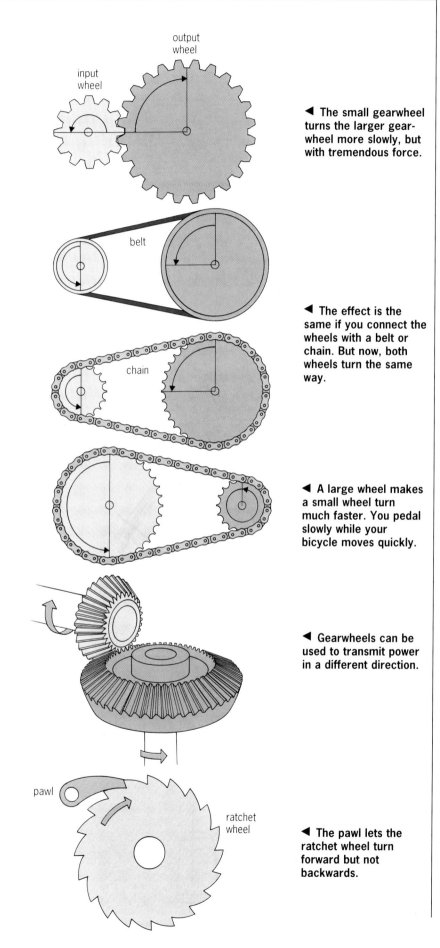

◀ The small gearwheel turns the larger gearwheel more slowly, but with tremendous force.

◀ The effect is the same if you connect the wheels with a belt or chain. But now, both wheels turn the same way.

◀ A large wheel makes a small wheel turn much faster. You pedal slowly while your bicycle moves quickly.

◀ Gearwheels can be used to transmit power in a different direction.

◀ The pawl lets the ratchet wheel turn forward but not backwards.

Time

Clocks and watches are used for telling the time. The mechanical sort with an hour hand and a minute hand were not made until the 17th century, but people have used timekeepers for many thousands of years.

The first clocks

Shadow sticks have been used since at least 3500 BC. The length and direction of the shadow cast by the Sun changes as the Sun moves through the sky. Sundials are a kind of shadow clock. You can read off the time at the point where the shadow falls on the dial. The stick that casts the shadow is called a 'gnomon'.

▶ On a sundial, the shadow changes position as the Sun moves across the sky. The position of the shadow tells you the time.

Water clocks (*clepsydras*) were used in Egypt from about 1400 BC. Time was measured by how long it took water to flow out of holes in a container. The Greeks and Romans made more complicated water clocks, like a cylinder into which water dripped from a reservoir. The time was read from a float.

▶ 16th century sand-glass. Turned so that all the sand was in the top half, it would take 15 minutes for the sand to flow through to the bottom.

Lamps, where the level of oil got lower as it burnt away steadily, and candles were used for centuries to tell the time. King Alfred the Great of Wessex is said to have used candles for timing in the 9th century.

Sand-glasses have been used from the Middle Ages. They are still often used as kitchen egg-timers. Large ones were made to time periods of half an hour or more.

Mechanical clocks

Galileo discovered that a swinging pendulum keeps regular time, but only used it to make a little instrument for doctors to count a patient's pulse beats. His son, Vincent, was the first to try and make a pendulum clock.

The first successful pendulum clock was made in 1657 in The Hague on instructions from the scientist Huygens. It can be seen in a museum in Leiden, Holland.

The first mechanical clocks were made in the 14th century. They did not have hands or a dial but made an alarm ring every hour. Striking clocks started to be put in public places in large cities in Europe. Soon, smaller versions were made for homes. The time was kept by a heavy bar, pivoted at the middle, that swung to and fro.

Pendulum clocks were invented in the 17th century. They were more accurate, because a swinging pendulum keeps regular time. To start with, the pendulums were short. The first long-case ('grandfather') clocks were made in 1670.

SCIENCE & MATERIALS

Watches

Pocket watches became possible after the invention in about 1500 of the mainspring (a tightly wound coil) to give the power. Wrist-watches started to get popular around 1900, when they were made mainly in France and Switzerland.

Modern clocks and watches

Mechanical clocks get their power from a weight that slowly falls or a spring that has to be wound up from time to time. In the 19th century, clocks driven by electric power were first made and by 1918 they could use the signal from mains electricity to keep time.

Today, many clocks and watches use the natural vibrations, 100,000 times per second, in a quartz crystal to keep time and their power comes from batteries. Even a small watch can be like a tiny computer, with a built-in alarm and stop-watch and the time shown by a digital (that is, in numbers) electronic display.

Atomic hydrogen masers are the most accurate timekeepers. They are accurate to within one second in 1,700,000 years.

Thomas Tompion (1638–1713) made the first watch to use a spring balance, an idea invented by the scientist Robert Hooke.

In 1772 John Harrison made the first chronometer, for keeping accurate time on long sea journeys. This would help mariners navigate.

The first electric clock was invented in 1843 by Alexander Bain, but it was not very reliable.

▲ Probably the oldest surviving working clock in the world is at Salisbury Cathedral in Wiltshire. It dates from 1386 and most of it is thought to be original. It has no face but signals the time by chimes.

▶ Modern watches can be bright, colourful and inexpensive. With electronics to make them work, they are also extremely accurate.

TELLING TIME

Watches that use hands and a dial to show the time are called **analogue** watches. This watch is reading a quarter to four, or 3.45. When you look at the hands it is easy to see that there are fifteen minutes to go before four o'clock.

Watches that use numbers to show the time are called **digital** watches. They keep time very accurately. This watch is also reading 3.45, but you cannot see at a glance how long it is until four o'clock. You have to work it out.

Metals

The Earth contains about 100 different elements, substances that are not mixtures of anything else. More than 70 per cent of these elements are metals such as gold, copper, tin and iron.

Pure metals

Pure metals are solids at room temperature. The one exception is mercury, a heavy, silver-coloured metal which is liquid. Mercury is shiny, and so too are all other metals. However, not all metals stay shiny when left in the air. Aluminium, lead, silver and many other metals eventually become dull and tarnish because elements such as oxygen in the air make new chemical compounds on the surface.

Metals, again with the exception of mercury, are strong. They can be bent without breaking. They can also be hammered or rolled into different shapes when hot, and pulled out to form wires.

▲ Gold nugget.

Metals are also good conductors of electricity and heat. Many metals make a ringing sound when struck.

Metals in nature

A few metals are found in the Earth in a pure state. These include platinum, copper, some silver and also gold, which may form lumps known as nuggets. We obtain most other metals from minerals known as ores. An ore is a mixture of a metal with other elements.

Mixing metals

Most metals that we see every day are in fact mixtures. These mixtures are known as alloys. An alloy is often quite different from the metals that make it up. Bronze, for example, is hard and strong. However, it is made up of copper and tin, two metals that are weak and soft. There are hundreds of different alloys, and more are invented every year.

Properties of metals

Most metals are very strong and are hard to stretch and bend.

Metals conduct heat and electricity well.

Most metals are shiny when newly made.

Metals are malleable, meaning that they can be hammered into thin sheets.

Metals are ductile, meaning that they can be drawn out into wires.

Many metals make a ringing sound when struck.

Common alloys

Brass contains copper and tin.

'Silver' coins are really made of copper and nickel.

Duralumin is an alloy of aluminium and copper. It is light and strong, and used in aircraft bodies.

Steel is a mixture of iron and carbon. Stainless steel also contains chromium, forming an alloy that does not rust.

Aluminium
Light and soft. Makes strong light alloys used for drinks cans, aircraft building, kitchen foil, high-voltage cables.

Copper
Good conductor of heat and electricity. Used for wires and for pipes used in plumbing.

Gold
Soft, very heavy and easily hammered into thin sheets. Does not corrode. Used for jewellery, reflective coatings.

Iron
Soft when pure, but very strong when made into steel. Rusts on exposure to air and moisture.

Lead
Soft and heavy. Used in sheets for waterproofing roofs. Poisonous.

Magnesium
Makes strong light alloys with aluminium and zinc used in aircraft and cars. Pure magnesium burns with a brilliant white flame and is used in fireworks.

Mercury
Liquid at room temperature. Heavy and poisonous. Used in switches, pesticides and thermometers.

Nickel
Does not tarnish or corrode easily and is magnetic. Used in alloys with iron and steel to make them stronger and more resistant to corrosion. Also used to make coins.

Platinum
Easily shaped. Does not corrode. Used in jewellery and as a catalyst in car exhausts to reduce pollution.

Silver
Used mainly in decorative objects and photography. Slowly tarnishes in air, becoming dull and finally black.

Tin
Does not corrode. Used mainly as a covering to prevent rust, and also mixed with lead to make solder.

Tungsten
Strong and hard. Used for light bulb filaments and in special steels to make cutting edges of saws and drills.

Uranium
Rare radioactive metal used to provide energy in nuclear reactors.

Zinc
Dull-coloured metal used to coat steel (galvanize) to prevent rust.

The discovery of metals

The first man-made metal objects date back to about 6000 BC. At that time gold and silver were made into bracelets and ornaments, while beaten copper was made into cooking pans and dishes.

In Egypt about 4000 BC men discovered how to smelt copper from copper ore. Later they learned that if they added tin the resulting bronze was much tougher, and did not rust or corrode. It could be used for cooking utensils, axe-heads and arrowheads.

The Bronze Age lasted from about 3500 BC to about 2000 BC, when iron smelting was invented in Anatolia (now Turkey).

Modern metal detectors

Today, there are many different ways of finding metal ores. Rocks which contain metals are much better conductors of electricity than other rocks, so electrical instruments can be used. Seismic instruments send shock waves through the rocks. The waves form special patterns when they pass through dense metal-bearing ores. Such dense rocks can also be found by instruments designed to measure gravity. Uranium minerals are radioactive, and can be located with a Geiger counter. Iron-containing minerals are magnetic.

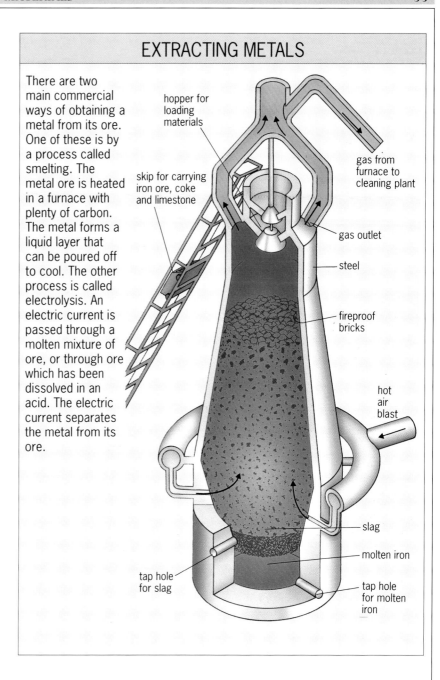

EXTRACTING METALS

There are two main commercial ways of obtaining a metal from its ore. One of these is by a process called smelting. The metal ore is heated in a furnace with plenty of carbon. The metal forms a liquid layer that can be poured off to cool. The other process is called electrolysis. An electric current is passed through a molten mixture of ore, or through ore which has been dissolved in an acid. The electric current separates the metal from its ore.

▲ A Bronze Age bowl made of bronze. Bronze is hard and tough, but it can be beaten into shape and carved into intricate designs.

Rust

Rust is a chemical substance called iron oxide. It forms a reddish-brown coating whenever iron or steel comes into contact with air and water. The rust falls off in flakes, leaving a fresh surface of metal open to the air. More rust forms, until eventually the metal is completely worn away.

Bicycle wheels and car trims are usually coated with chromium, which does not rust. But if the chrome gets scratched or chipped, air and water can reach the metal below and rust starts to form.

Rust protecters

'Galvanized' metal buckets and corrugated iron roofs are coated in the metal zinc.

Cans used for storing food are made of steel covered with a thin coat of tin.

Iron and steel are coated with oil, grease or paint.

Plastics

► A collection of objects made of different types of plastic materials.

The first plastic ever produced was Parkesine. It was made by a British chemist, Alexander Parkes, in 1862. It felt and looked like ivory.

Many plastics have 'poly' in their name: for example polyethene (more commonly known as polythene). *Poly* comes from a Greek word meaning 'many'. *Polyethene* means 'many molecules of ethene joined together'.

Plastics do not occur naturally but must be manufactured. Most plastics are made from chemicals found in oil, although a few come from wood, coal and natural gas. Common types include polythene, polystyrene, PVC and nylon.

Plastics do not rot like wood, or rust like iron and steel. They are light in weight and can be made in almost any shape or colour. Most do not allow electricity to pass through them. This means that they can be used to cover wires, plugs and other electrical items. Some plastics are hard. Others are soft and stretchy. Many are transparent. Some can be filled with tiny gas bubbles to make plastic foam. Some can be drawn out into fine fibres and woven into cloth, and some can be used to make paints and glues.

Making plastics

Plastics are made by treating chemicals made from oil (or another starting material) with more chemicals, heat and pressure. The result is a material whose molecules link together to form long chains. Materials like plastics, which have long-chain molecules, are called **polymers**. It is this chain-like structure which gives plastics their special properties.

When newly-made plastics are still hot, they are soft and easy to mould. In fact, the word 'plastic' means mouldable. There are two main types of plastics. **Thermoplastics** set solid when first cooled, but can be remelted and moulded over and over again. The plastics used for buckets and washing-up bowls are like this. **Thermosets** cannot be remelted once they have set. The plastics used for saucepan handles are like this.

Shaping plastics

There are various ways of shaping plastics. Plastic bottles are **blow moulded**. Air pressure is used to push a 'bubble' of hot, soft plastic against the inside of a mould. Vacuum forming uses a similar idea: a sheet of soft plastic is sucked against a mould to form a large shape like a bath or a sink. Combs and plastic toys are **injection moulded**. Melted plastic is forced into a mould. Curtain rails, pipes and wire insulation are formed by **extrusion**. The melted plastic is squeezed out through a specially-shaped opening rather like toothpaste coming out of a tube. In **calendering**, rollers are used to squeeze plastic into thin sheets.

Fibres from plastics

Many synthetic fabrics are made from plastic fibres. These include nylon, polyester and 'acrylic' materials such as Terylene, Dacron and Acrilan. Fibres are made from plastics by melting or dissolving the plastics and then forcing them through tiny tubes called spinnerets to make fine threads. Polyester fibres are used in blouses, dresses, shirts and other lightweight clothes. Acrylic fibres feel soft like wool. They are used to make blankets and winter clothes.

Making plastic disappear

Plastic is used to make so many things that it makes up a large part of the rubbish we throw away every day. About 100 million tonnes of plastics are made each year. Plastic rubbish does not break down easily like the vegetable scraps and old paper.

Scientists are now finding ways to make new plastics that will break down in rubbish tips. Some are broken down by sunlight, while others can be eaten by bacteria. A remarkable new plastic called Biopol is made from sugar by special bacteria and used for things like shampoo bottles. Bacteria in the soil can break it down again.

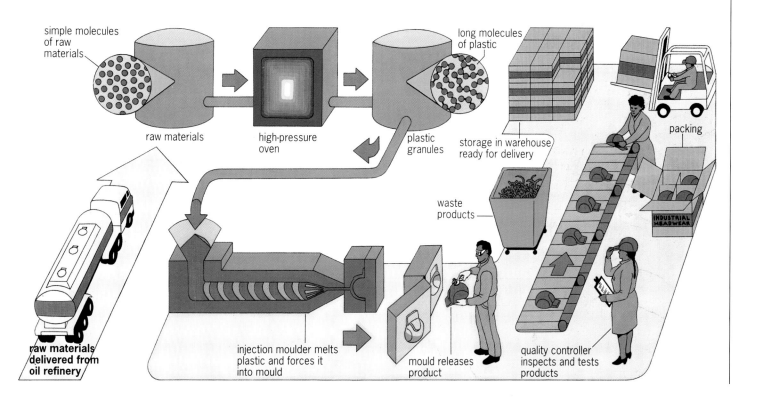

▼ A plastics factory. Using raw materials a range of plastic products are produced.

Space Exploration

Probes to the planets

Venus
Mariner 2, 1962, first fly-by
Venera series, 1967 to 1980s, landed on surface
Pioneer Venus 1 & 2, 1978, orbited and probes entered atmosphere

Mars
Mariner 4, 1964, first pictures
Mariner 9, 1971, orbited
Viking 1 & 2, 1975, orbited and landed
Phobos 1 & 2, 1988, to Martian moons

Jupiter
Pioneer 10 & 11, 1972–1973, fly-by
Voyager 1 & 2, launched 1977; flew past planet and moons in 1979

Mercury
Mariner 10, 1973, 10,000 pictures

Saturn
Pioneer 11, flew by in 1979
Voyager 1 & 2, flew past in 1980–1981

Uranus
Voyager 2, flew past in 1986

Neptune
Voyager 2, flew past in 1989

▶ Space Shuttle lifts off.
Weight at launch
Including boosters and fuel tank: 2,000 tonnes
Height at launch
56 m
Length of shuttle orbiter
37 m
Wingspan
24 m
Weight
84 tonnes
Total power at launch
The same as 140 Jumbo Jet aircraft

On Earth we are surrounded by air, but this gradually gets thinner as it gets further from the ground. Beyond the air there is almost empty space stretching out between the Sun, the Moon, the planets and the stars. Space is not easy to explore because everything is so far away. Astronauts have visited only our nearest neighbour, the Moon. They plan to visit the planet Mars but it will take about six months to get there. Robot spacecraft can explore the more distant planets but it would take them thousands of years to get to even the nearest star.

Going into space

To explore space we have to escape from the pull of gravity, which holds us down onto the Earth. We can do this by travelling very fast. You need a speed of 40,000 km/h (25,000 mph), about 20 times as fast as Concorde, to escape completely. This is called the Earth's escape velocity. Satellites which go into orbit circling round the Earth need to go only about 12 times as fast as Concorde. Only powerful rockets can launch astronauts or satellites into space, because they can reach these speeds and a rocket engine will work in space, unlike aircraft engines.

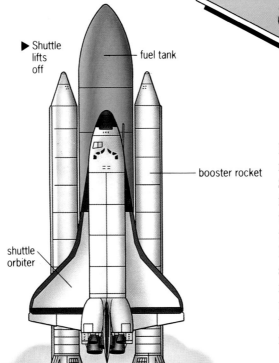

▲ Fuel tank falls away.

◀ Boosters fall away.

▶ Shuttle lifts off
— fuel tank
— booster rocket
shuttle orbiter

Space planes

Rockets are very expensive to build and most can be used only once. The Space Shuttle is the first space plane which can fly into space many times. The only part that is new for each flight is the huge fuel tank. The two booster rockets which help lift the Shuttle into space fall back into the sea when their fuel runs out and are recovered by ships. The Shuttle takes off upwards like a rocket, but when it returns to Earth it lands like a giant glider on a runway. Space planes of the future may be completely re-usable, taking off and landing on a runway like an aircraft.

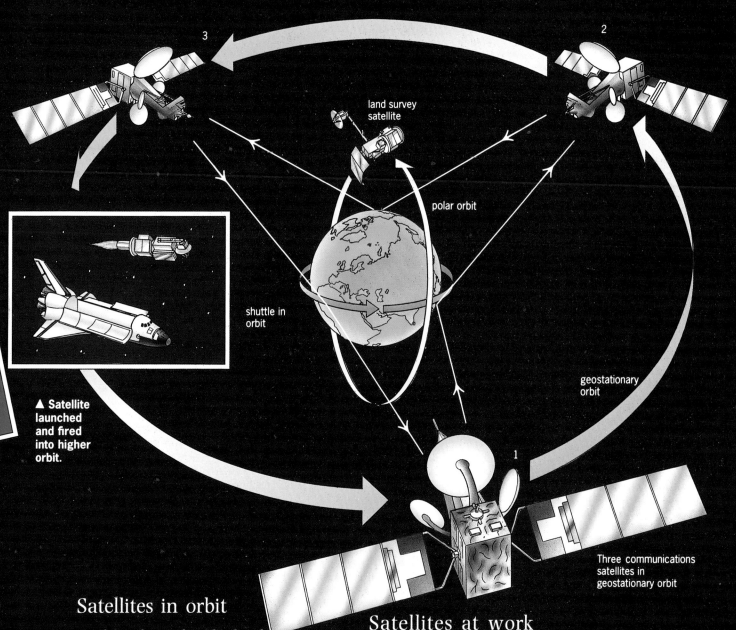

land survey satellite
polar orbit
shuttle in orbit
geostationary orbit
Three communications satellites in geostationary orbit

▲ Satellite launched and fired into higher orbit.

Satellites in orbit

Racing round Earth are hundreds of satellites doing many different jobs. Their orbits have different heights and directions depending on what the satellite does. Communications satellites which send TV and telephone signals across the oceans use the high 'geostationary orbit', about 35,000 km (22,000 miles) above the Equator. They seem to hover above the same point on Earth because they circle at the same rate as the Earth spins round. Lower down, other satellites watch the Earth or the weather, circling north and south while the Earth turns beneath them.

Satellites at work

Although satellites have different jobs to do, they all need the same basic parts. Most get their power from panels of solar cells which turn sunlight into electricity. They have small gas jets to move them or turn them round, and sensors to check that they are facing the right way. Radio aerials keep them in touch with engineers on Earth. It is difficult to repair satellites in space, so they are tested very thoroughly before launch to make sure they will not break down.

▼ Shuttle glides back to Earth and lands.

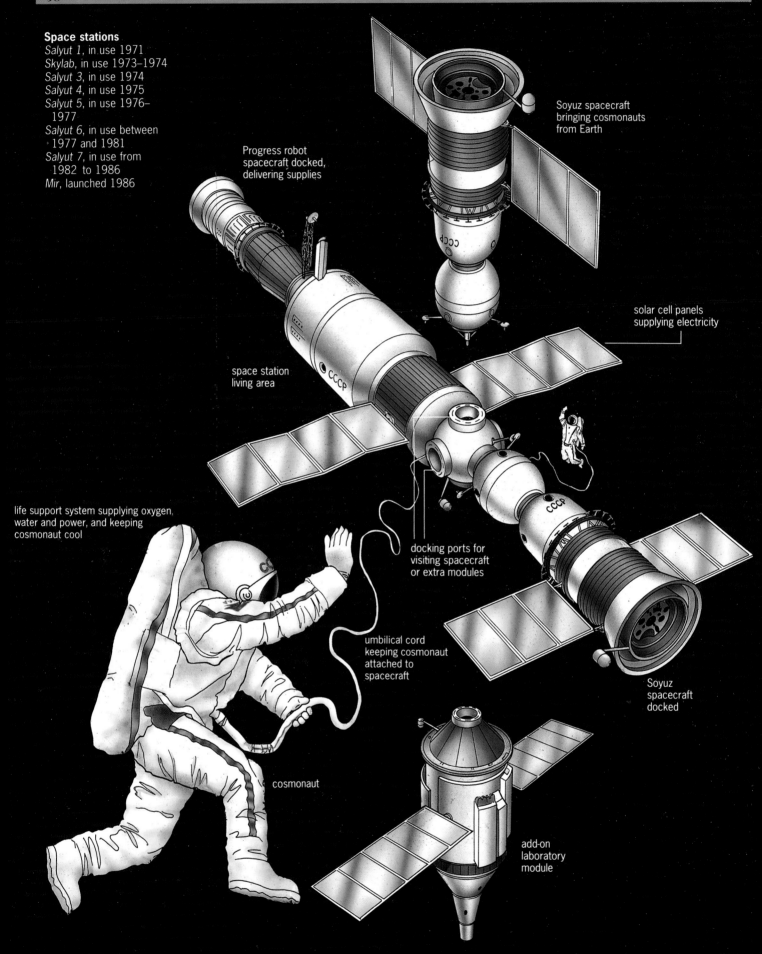

Space probes

The robot spacecraft sent to explore other planets are called probes. Some fly past the planet, while others circle round it, sending information back to Earth. The *Voyager* spacecraft flew past Jupiter, Saturn, Uranus and Neptune, sending us beautiful close-up pictures of the planets and many of their moons. Orbiting probes have photographed almost all the surface of Mars. The ones circling Venus used radar to map the surface hidden by its thick clouds. Pluto is the only planet that has not been visited.

Robot explorers

We can get a closer look at the Moon or planets by landing robot explorers on them. *Lunokhod 1* and *2* were remote-controlled cars which explored the Moon, crawling slowly over its surface testing its soil. Two *Viking* spacecraft landed on Mars to study the soil and weather. They searched for signs of life on Mars, with confusing results. Spacecraft which land on Venus have to be very tough to survive the heat and the atmosphere, which presses down very hard. Some have lasted long enough to send back pictures of the surface.

Man on the Moon

In 1969 the giant *Saturn 5* rocket launched three astronauts towards the Moon. They travelled for over three days in a small spacecraft called the Apollo Command Module. Two of the astronauts landed on the Moon in the Lunar Module, leaving the third circling the Moon. After almost a day the Lunar Module took off and met the Command Module, which took them all back to Earth. Five more visits followed, when the astronauts explored the Moon, collecting moonrock, using a moon car called the Lunar Rover on three visits.

Space stations

A space station is a home in space, circling round the Earth, where astronauts can live and work. It contains everything they need, including food, water and air, which must all be brought up from the Earth. It must be strong enough to hold in the air for the astronauts to breathe and to protect them from radiation and the dust speeding through space. Large space stations where people can live for many months are built in space by adding extra sections sent up from Earth.

Returning to Earth

The most dangerous parts of any space flight are the launch and the return to Earth. A returning spacecraft meets the blanket of air surrounding the Earth at very high speed. It rubs against the air, getting so hot that the outside of the spacecraft glows red. This would melt the spacecraft and kill the astronauts if they were not protected. Early spacecraft had a thick outer layer which burned away, but the Space Shuttle has special re-usable heat-proof tiles.

Junk in space

About 3.5 million pieces of rubbish are orbiting the Earth in space today. Worn-out satellites and rockets, tools lost by astronauts and even tiny flecks of paint continue to go round and round. They are travelling at such high speeds that even a tiny chip of metal 0.5 mm wide could pierce the spacesuit of an astronaut and kill him. A piece a few centimetres across could destroy a space station. Space junk poses a big problem for future space travellers.

Further into space

A great deal of space exploration can be done right here on Earth.

The Earth's atmosphere tends to blur radiation passing through it, and there is also interference from Earth sources of light and radio waves. Astronomers get a clearer picture by putting their telescopes on satellites.

Apollo missions to the Moon
Apollo 11, July 1969, first men on Moon
Apollo 12, November 1969, 32 hours on Moon
Apollo 13, April 1970, no landing because explosion on spacecraft, astronauts returned
Apollo 14, February 1971, visited highlands
Apollo 15, July 1971, first used Lunar Rover
Apollo 16, April 1972, drove 27 km (17 miles) on Moon
Apollo 17, December 1972, last Moon visit

Future space stations may be like giant, slowly spinning wheels. The spinning will hold astronauts against the outer wall, just as if gravity were acting.

The largest orbiting telescope is the Edwin Hubble Space Telescope. It is over 13m long, has a reflector 240cm across, and weighs 11 tonnes.

Space is measured in light years. 1 light year is almost 10 million million kilometres.

Astronauts

Astronauts are people who leave the Earth and travel into space. In Russia they are known are cosmonauts. The first was Yuri Gagarin, who orbited once round the Earth in his spacecraft on 12 April 1961.

The Moon is the furthest that any astronauts have travelled from Earth so far. Neil Armstrong and Edwin 'Buzz' Aldrin of the USA were the first to land there on 20 July 1972. Ten other astronauts have travelled to the Moon since, the last in 1972. Missions to Mars are now being planned.

▼ Astronaut Guion Bluford about to be lowered into the training pool in the Johnson Space Centre. Working at the bottom of this deep pool simulates a weightless environment.

Living in space

One of the biggest problems for astronauts is weightlessness. In a space station everything floats around. If astronauts want to stay in one place they have to fix themselves to something. They sleep strapped into a sleeping bag. The food is normal but it is made sticky so it sticks to a spoon, and drinks are in containers with a straw to prevent them from floating around as a ball of liquid.

Astronauts wash inside a closed shower-bag. This prevents the water floating all round the spacecraft. They also have to use a special toilet. It has a fan which sets up a draught of air so that liquid and solid waste is sucked into the toilet. Weightlessness also affects astronauts' bodies, so they have to exercise hard while they are in space to keep fit.

Astronaut training

Before they go into space, astronauts have to practise all the things they will need to do when they are there. To get used to weightlessness, they fly in an aircraft which climbs and then dives in a special curved path so that the astronauts can float about in their cabin. To practise control of their spacecraft, they use a full-size working model, called a simulator, which stays on the ground.

Working in space

Astronauts work hard in their space stations, studying the Earth below, the distant stars and galaxies or the space around them. They measure the effects of weightlessness on themselves, on growing plants and on small creatures such as spiders and bees. Some crystals and alloys (mixtures of metals) are very difficult to

SPACE

▲ Astronaut walking in space.

make on Earth because gravity churns up the melted materials they are made from. In space astronauts can make crystals, alloys and also some medicines for use on Earth.

Spacesuits

When astronauts leave their spacecraft they must wear a spacesuit, with its helmet and gloves. Its many layers of material protect against radiation and dust. Inside is an inflated layer (a bit like a balloon) which presses against the body. Without this pressure the blood would boil. Next to the skin are tubes with cooling water to take away trapped body heat to a backpack, which also carries a power supply and oxygen gas to breathe.

When they take a space-walk, astronauts are attached to a strong line to prevent them floating away into space. On the Moon, astronauts use a special buggy called a Lunar Rover, which runs on batteries.

A risky business

Longer missions into space pose problems for astronauts. The intense radiation that is found in outer space can cause cancer. Space scientists must find a material that will not let the dangerous rays into the spacecraft.

Weightlessness has several bad effects on the human body. It upsets our balance, causing space sickness, but this wears off after a time. Bones become thin and muscles waste away in space. Bacteria that cause disease multiply faster when they are weightless, but the body's defences build up more slowly.

FAMOUS ASTRONAUTS

▲ **Neil Armstrong** was born in 1930 in Wapakonetz, Ohio, USA. He learned to fly at the age of 16, and became a naval pilot.

Later in the 1950s he became a test pilot for NASA (National Aeronautics and Space Administration) before joining the US space programme in 1962. His first mission in space, on *Gemini 8* in 1966, ended earlier than expected when he had to make an emergency landing in the Pacific Ocean.

In 1969 he joined astronauts Aldrin and Collins on the *Apollo 11* mission, and on July 20 he became the first person to walk on the Moon. As he stepped off the lunar landing module, he said, 'That's one small step for man, one giant leap for mankind.'

First person in space
On 12 April 1961 Yuri Gagarin of the USSR circled the Earth once at a height of about 400 km (250 miles) and a speed of 29,000 km/h (18,000 mph). The whole trip lasted 108 minutes.

First 'space walk'
Alexei Leonov left the spacecraft for 12 minutes on 18 March 1965.

Longest space stay
In December 1987, Yuri Romanenko of the USSR completed a stay of 326 days aboard the MIR space station.

Yuri Gagarin was born in 1934 in Gzhatsk, Russia, the son of a poor farmer. The town has now been renamed Gagarin. He was the first human to travel in space. Gagarin left school to train as a steel worker. At college he joined a flying club, and soon joined the air force and began to fly fighter planes. Gagarin was a small man, which helped him as an astronaut, since the first manned spacecraft, *Vostok*, was very small. He died in a plane crash in 1968 while testing a new plane.

▼ **Valentina Tereshkova** was the first woman in space. She was born in 1937 near Yaroslavl, Russia, and lived on a farm. She became a factory worker, but took up sky diving as a hobby, and joined the space mission in 1962.

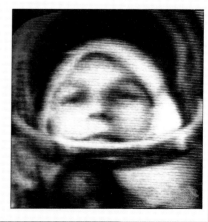

Rockets

▶ The *Ariane* rocket can launch 2 or 3 satellites into orbit at the same time.

Saturn 5 rocket
Height
110 m (twice as tall as Nelson's Column, London)
Weight
At lift-off 3,000 tonnes; 90 per cent of this was fuel
Power
First-stage engines: the same as 160 Jumbo Jet engines

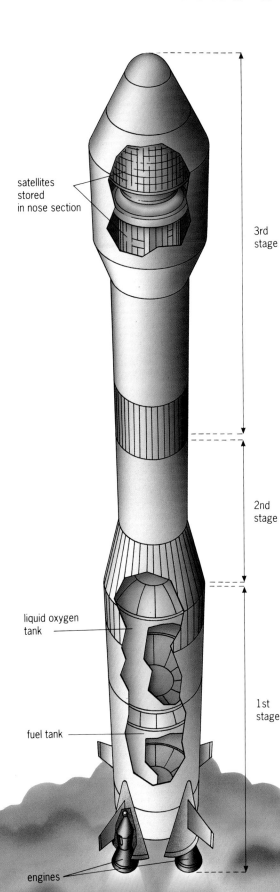

The huge rockets that send satellites into space and the beautiful firework rockets work in exactly the same way. They are full of fuel which burns to make a lot of hot gas. This expands rapidly. The force of the expansion pushes the gases downwards and the rocket upwards. You can see how this works if you blow up a balloon and let it go. The air rushes out and the balloon shoots away. It goes all over the place because it is soft and the air can escape in any direction. Space rockets go in the right direction because nozzles direct and control the gases.

Space rockets

Space rockets have to be very powerful to escape from the strong pull of the Earth's gravity and send a spacecraft or satellite into space. They usually have several stages; they are really two or three rockets stacked on top of each other. The first stage, at the bottom, lifts the rocket off the ground, thrusting it up until the fuel runs out. It then falls away and the second-stage engines take over, and so on. So the rocket does not carry unnecessary weight into space. Most rockets are very expensive because they are used only once, but with the Space Shuttle only the fuel tank is wasted. The rest returns to Earth and is used again.

Rocket fuels

In some space rockets the fuel is solid, like rubber. These rockets are often boosters, extra rockets fixed to the side of the main rocket. However, most space rockets use the more powerful liquid fuels in huge tanks inside the rocket. The fuel will not burn without oxygen so a second tank carries an oxygen supply. Rockets are the only engines that will work in space where there is no air to supply oxygen. The jet engines on aircraft use oxygen from the air so they will not work in space.

Flashback

The first rockets were made nearly 1,000 years ago in China. They were like our firework rockets and were used in battles, fixed to arrows. Rockets were later used to save lives, as distress signals. In 1926 Robert Goddard launched the first rocket with an engine that burned liquid fuel. The German *V-2* war rocket, made in 1942, was the first rocket powerful enough to reach space, but it was not until 1957 that a Soviet rocket launched the first satellite into space. In 1969 the giant American *Saturn 5* rocket launched the first astronauts to land on the Moon.

The most powerful rocket ever launched, *Saturn 5*, weighed up to 3447 tonnes. During take-off, it consumed 13·6 tonnes of fuel per second for 2½ minutes – 2042 tonnes in all.

Pioneer 10, in 1972, was the first rocket to get up sufficient speed to leave the Solar System. It left the Earth at 51,682 km/h.

The fastest rocket-propelled object ever is the solar probe, *Helios B*, which reaches 252,800 km/h at certain stages of its solar orbit.

WERNHER VON BRAUN

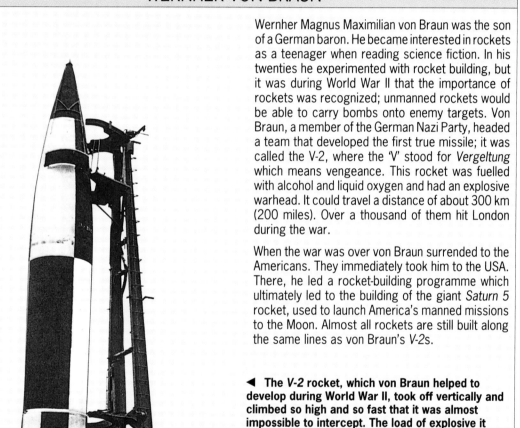

Wernher Magnus Maximilian von Braun was the son of a German baron. He became interested in rockets as a teenager when reading science fiction. In his twenties he experimented with rocket building, but it was during World War II that the importance of rockets was recognized; unmanned rockets would be able to carry bombs onto enemy targets. Von Braun, a member of the German Nazi Party, headed a team that developed the first true missile; it was called the V-2, where the 'V' stood for *Vergeltung* which means vengeance. This rocket was fuelled with alcohol and liquid oxygen and had an explosive warhead. It could travel a distance of about 300 km (200 miles). Over a thousand of them hit London during the war.

When the war was over von Braun surrendered to the Americans. They immediately took him to the USA. There, he led a rocket-building programme which ultimately led to the building of the giant *Saturn 5* rocket, used to launch America's manned missions to the Moon. Almost all rockets are still built along the same lines as von Braun's *V-2s*.

◀ **The *V-2* rocket, which von Braun helped to develop during World War II, took off vertically and climbed so high and so fast that it was almost impossible to intercept. The load of explosive it carried was small, but its maximum speed was an astonishing 5,470 km/hour (3,400 mph). Germany bombarded London and Antwerp with many hundreds of V-2s during the last months of the war.**

The Moon

Average distance from Earth
384,400 km (239,000 miles)
Diameter
3,476 km (2,160 miles)
Mass
1/81 the Earth's mass
Volume
1/50 the Earth's volume
Surface gravity
0.165 the Earth's gravity
Time between new Moons
29.53 days
Time to rotate on axis
27.32 days (relative to the stars)
Surface temperature
120°C maximum to –153°C at night
First successful Moon probe
Luna 1 (January 1959)
First craft to soft-land
Luna 9 (January 1966)
First manned landing
Apollo 11 (July 1969)
First man on Moon
Neil Armstrong (21 July 1969)

▼ The American lunar module 'Falcon' and its Commander, James B. Irwin, on the Moon in 1972. The special Lunar Roving Vehicle is standing beside it.

The Moon is Earth's natural satellite. It is made of rock and is like a small planet. In fact it is not much smaller than the planet Mercury.

Every year, the Earth and the Moon make a trip together round the Sun. While that is happening, the Moon is going round the Earth in orbit as well, but it always keeps the same side pointing towards us. Spacecraft have taken photographs of the 'back' of the Moon and they show that it is quite similar to the side we can see.

Just like the Sun, the Moon rises and sets each day because the Earth is turning on its axis. It takes the Moon one month to make a complete orbit round the Earth.

What it is like on the Moon

We know a lot about the Moon because huge numbers of close-up photographs have been taken by spacecraft, and astronauts have travelled there. It is a totally lifeless place with no air and no water. That means there is no wind or weather to change the surface. The footprints the astronauts made will remain for millions of years! The pull of gravity at the Moon's surface is only one-sixth as strong as on Earth, which is not enough to keep an atmosphere. The astronauts felt much lighter there and could jump easily, even in their heavy spacesuits.

PHASES OF THE MOON

The Moon does not give out any light of its own. It shines only because it reflects a little bit of the sunlight that falls on it. Only the half that is facing the Sun is lit up. As the Moon moves in its orbit around the Earth, the Earth casts a shadow on it, so at times it appears crescent-shaped. The different shapes are called the Moon's phases.

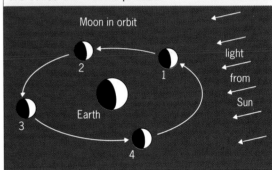

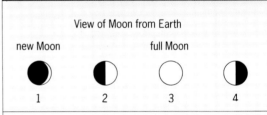

This shows how the phases of the Moon change. As the Moon travels round Earth, its shape appears to change as it moves in and out of Earth's shadow.

Nearly everywhere there are craters of every size from microscopically small up to about 250 km (150 miles) across. They were made hundreds of millions of years ago by meteorites that crashed into the Moon not long after it formed. Some large patches on the Moon's surface look darker than the rest and do not have so many craters. They were created by hot molten rock from inside the Moon that flooded into enormous craters, then solidified as it went cold. If you look at the Moon through a telescope or a pair of binoculars, you will be able to see the dark areas more clearly, and lots of craters. They show up best near the dividing line between the lit and unlit parts of the Moon, where the crater walls and mountains cast long shadows.

Solar System

Our Sun has travelling around it a family of planets, some of them with moons, and lots of other smaller objects such as comets and meteor streams. The Sun and all things in orbit around it make up the Solar System, which is over 12,000 million km (7,500 million miles) across.

The largest objects in the Solar System, apart from the Sun, are the nine planets: Mercury, Venus, Earth, Mars, Jupiter, Saturn, Uranus, Neptune and Pluto. All except Mercury and Venus have at least one moon and more than fifty moons are known altogether. There are thousands of minor planets, hundreds of comets and streams of dust and pieces of rock. Astronomers believe that these things are left over from when the Sun formed from a cloud of gas about 5,000 million years ago. The Solar System stays together because of the strong pull of the Sun's gravity.

The paths of the planets

The orbits of the planets round the Sun are not circular but have the squashed oval shape called an ellipse. None of them is tilted very much to the others and so the Solar System has the shape of a flat disc. Pluto is an exception. Its elliptical orbit crosses just inside Neptune's and is tilted at 17°. The way gravity acts means that the further a planet is from the Sun, the longer the period of time it takes to complete an orbit.

The spaces between the planets are huge compared to their sizes. If the Sun were a football, the Earth would be this big ● and 30 m (98 ft) away!

The word 'solar' means 'to do with the Sun'. It comes from *sol*, the Latin word for Sun.

▼ The major planets in order from the Sun. The planets are drawn to scale, but the distances between them are not to scale.

	Pluto	Neptune	Uranus	Saturn	Jupiter	Mars	Earth	Venus	Mercury
Distance from the Sun in million km	5,900	4,497	2,870	1,427	778	228	150	108	58
Time to orbit Sun in days	90,502	60,275	30,660	10,767	4,343	687	365	225	88

Planets

▼ Three separate photographs, taken through different coloured filters by the spacecraft *Voyager 1* in 1980, were used to construct this composite picture of Saturn. The colours have been made brighter so that the bands of clouds show more clearly as well as the rings.

Earth is one of nine major planets that orbit round the Sun in the Solar System. The names of the planets, going in order outwards from the Sun, are: Mercury, Venus, Earth, Mars, Jupiter, Saturn, Uranus, Neptune and Pluto. So far, no planets going round other stars have been found for certain.

Planets are different from stars because they do not give out any light of their own. They shine because they reflect sunlight. Some planets are balls of rock like the Earth. Jupiter, Saturn, Uranus and Neptune are giant balls of gas.

In 1977, astronomers discovered that Uranus also has a set of narrow rings. They are too faint to be seen and were only found when Uranus crossed in front of a star and the rings cut out the starlight for a few moments. Jupiter has a faint ring, discovered by the space probe *Voyager 1* in 1979. In 1989, *Voyager 2* found three rings around Neptune.

Natural satellites

Many of the planets have natural satellites (moons) orbiting round them. The largest ones are similar in size to Mercury but there are many that are just small chunks of rock.

The biggest moons are: the Earth's; the four around Jupiter discovered by Galileo and called Io, Europa, Ganymede and Callisto; Saturn's moon Titan; and Triton, which orbits Neptune. Mars has two tiny moons called Phobos and Deimos. They are Greek words for 'fear' and 'terror'. Jupiter has at least sixteen moons altogether; Saturn twenty, Uranus fifteen, Neptune eight and Pluto one, called Charon.

Mercury

Mercury is a small rocky planet 4,900 km (3,000 miles) across and 58 million km (36 million miles) from the Sun. It makes a circuit of the Sun in 88 days. Mercury spins round every 59 days. On the side facing the Sun, the temperature reaches 400°C (752°F) but at night it falls as low as -170°C (-274°F) because there is no atmosphere to trap the heat.

Rings around the planets

Saturn is circled by a beautiful system of rings, made of countless individual particles of dust and pieces of rock in a band less than 200 m (650ft) thick around the planet. Galileo was the first person to see the rings when he looked through his telescope in 1610. They are easy to see with a small telescope.

▶ This image of Mercury is a mosaic of pictures from the *Mariner 10* spacecraft, taken from a distance of around 210,000 km (130,000 miles). The surface is a bit like the Moon's, with lots of craters.

Venus

Venus shines brighter and gets nearer to Earth than any other planet in the sky. It is a rocky planet 108 million km (67 million miles) from the Sun. At 12,100 km (7,500 miles), its diameter is just a little less than Earth's. Venus has a thick atmosphere of carbon dioxide, which traps heat like a greenhouse and contains dense clouds. At 500°C (932°F), the surface is even hotter than Mercury's. It is impossible to see the surface through the clouds, but spacecraft have made maps by using radar. These show mountains, plains and craters.

Earth

There are many ways in which planet Earth is unique in the Solar System. Perhaps the most obvious one is that it has life. Life, as we know it, needs water, and three-quarters of the Earth's surface is covered by oceans. Unlike the other rocky planets, the inside of the Earth is very active. Movements of the Earth's crust and volcanoes create new mountains, while rain and wind wear down the land.

Mars

Mars is a small, rocky planet 6,800 km (4,200 miles) across and 229 million km from the Sun. Its surface shows craters and several large, extinct volcanoes. No water flows here and the atmosphere of carbon dioxide is very thin. Like Earth, Mars has seasons. During a Martian year, which is nearly twice as long as an Earth year, polar ice-caps made of frozen carbon dioxide grow in winter and shrink again in summer.

The minor planets

In the gap between the orbits of Mars and Jupiter, thousands of pieces of rock circle the Sun. These are the minor planets or asteroids. The largest one, Ceres, was discovered first, in 1801. It is about 940 km (590 miles) across. Most asteroids have uneven shapes. Astronomers think they are small pieces left over from when the Solar System formed.

◀ What the planet Earth looks like from space. This is how *Apollo 17* astronauts saw it in 1972. White swirls are clouds and the blue is the oceans. Antarctica is at the bottom and Africa near the top left.

The Earth's diameter is 12,750 km (7,920 miles). Its distance from the Sun is 150 million km (93 million miles).

◀ Part of the planet Mars. The three spots in a line are the giant cones of extinct volcanoes. The long dark gash is an enormous canyon, called the Mariner Valley.

▼ The surface of Mars is like an orangey-red desert of rocks and dust. This view was taken by the *Viking 2* spacecraft which landed on Mars in 1976.

Jupiter and its moons

Jupiter, the giant of the Solar System, is 778 million km (483 million miles) from the Sun and 142,800 km (88,700 miles) across at its equator. It could swallow up more than a thousand Earths and has more material in it than all the rest of the planets together. It is the planet that shines brightest after Venus.

Coloured bands of clouds swirl around as Jupiter spins on its axis in less than ten hours. Unlike the rocky planets, it is a great ball of gas, mainly hydrogen, and does not have a solid surface under the clouds. The largest cloud feature is the Great Red Spot, which has existed for at least 140 years.

The four largest moons of Jupiter can be seen easily in a small telescope as small points of light near the planet. The pattern they make changes from day to day as each circles Jupiter in its own time. The *Voyager* space probes took close-up photographs in 1979 and showed for the first time what they are really like.

Io is the strangest. It has several active volcanoes that pour out sulphur, and colour the surface a mixture of yellow, orange and red. Callisto has the most heavily cratered surface in the Solar System but Ganymede has craters larger than any on Callisto. Europa's smooth icy surface is criss-crossed by numerous mysterious dark lines.

▶ The planet Jupiter and its four biggest moons, Io, Europa, Ganymede and Callisto. Separate views from the *Voyager* spacecraft have been put together to make this imaginary scene.

Uranus has a diameter of 52,400 km (33,000 miles).
It is 2,870 million km (1,783 million miles) from the Sun.

Neptune has a diameter of 48,600 km (30,000 miles).
It is 4,497 million km (2,794 million miles) from the Sun.

Saturn

Saturn is like a smaller version of Jupiter in many ways, a gas ball 120,000 km (75,000 miles) across at its equator. The main differences are the cloud bands, which are much fainter, and, of course, the amazing rings.

Saturn orbits the Sun at a distance of 1,427 million km (887 million miles), taking 29 years to make a complete trip. The rings are tilted to Saturn's path. As a result, we see the rings at different angles at different times.

Saturn's largest moon, Titan, is the only moon in the Solar System that has an atmosphere. Instruments on the *Voyager* space probe found that it is mainly made of the gas nitrogen. Thick reddish clouds make it impossible to see Titan's rocky surface.

Uranus, Neptune and Pluto

The planets from Mercury out to Saturn are all bright enough to be seen easily by eye alone. Uranus was the first planet to be discovered with the help of a telescope. It was found accidentally by William Herschel in 1781.

After a few years, astronomers noticed that Uranus was not keeping to the path they expected. They suspected that the gravity of an undiscovered planet might be pulling it. Two mathematicians set about working out where the planet should be. Their predictions resulted in the discovery of Neptune in 1846. Some astronomers were convinced that there might be yet another planet. They kept searching, but Pluto was not found until 1930.

Uranus and Neptune are similar gas-ball planets, both about four times larger than the Earth. A strange feature of Uranus is the way it rotates round an axis that is tipped right over at an angle of 98 degrees, so that the planet seems to be lying on its side. Neptune has a large moon, Triton, about the same size as Earth's, and at least seven others.

▲ An imaginary view of Saturn with five of its moons, made up from real pictures returned by the *Voyager* spacecraft. The one shown biggest is Dione. Then, going clockwise, the others are: Enceladus, Rhea, Mimas and Tethys. The starry background is a painting.

Pluto, the most distant planet, is a small, frozen planet about the size of our Moon. It has its own moon, Charon. Pluto's oval orbit takes it round the Sun once in 248 years and its distance from the Sun ranges between 7,400 and 4,400 million km (4,600 and 2,700 million miles). At this distance, it receives very little heat and the temperature must be about $-230°C$ ($-382°F$).

PLANET SPOTTING

Venus, Mars, Jupiter and Saturn all get very bright in the sky and are easy to see. You can usually tell a planet because it shines with a steady light and does not twinkle so much as the stars. The planets cannot be seen in the night sky all the time because of the way they travel round the Sun. You can find out when to spot the planets by looking out for the monthly sky charts in newspapers or astronomy magazines.

A really brilliant object in the western sky in the evening or in the east in the morning is sure to be Venus. Mars looks distinctly red and Jupiter and Saturn look yellowish. If you have a pair of binoculars, try looking at a planet through them and drawing what you can see.

Sun

Distance from the Earth 149,600,000 km (93 million miles)
Equatorial diameter 1,392,000 km (865,000 miles)
Rotation period 27¼ days (as seen from Earth)
Mass 332,946 x Earth
Volume 1,303,946 x Earth
Surface temperature 6,000°C
Core temperature 16 million°C

The Sun is an ordinary star, like the ones you can see at night, but for us and all the things that live on the Earth it is very special. Without the heat and light energy from the Sun, there would be no life. Even things we get from the ground and burn to get energy, such as coal, gas and oil, are the remains of plants and animals that grew in sunlight millions of years ago. People have always recognized how important the Sun is and they often worshipped it as a god.

Energy from the Sun

The Sun is a giant ball of hot gas, 150 million km (93 million miles) away. It measures as much across as 109 Earths side-by-side and could hold more than a million Earths inside its volume. Near the outside of the Sun the temperature is about 6,000°C but in the centre it is more like 16 million °C. In the Sun's hot core, enormous amounts of energy in the form of heat and light are produced by a process called nuclear fusion. This energy makes the Sun shine.

As well as heat and light, the Sun gives out X-rays and ultraviolet rays that are harmful to life. Most of these rays get soaked up in our atmosphere and do not affect us. Sunlight is very strong and **no one should ever stare at the Sun or look at it through any kind of magnifier, binoculars or telescope**. Your eyesight would be badly affected or you could even be blinded. Even with dark sunglasses or film it could still be dangerous to look at the Sun, so it is best not to risk it at all. Astronomers study the Sun safely by looking at it with special instruments.

The Sun gets 4,000 million tonnes lighter every second as it generates nuclear energy from hydrogen. It takes light 8·3 minutes to travel from the Sun to the Earth.

The Sun is 5,000 million years old.

In 5,000 million years the Sun will become so bright that it will be impossible for anything to live on our planet.

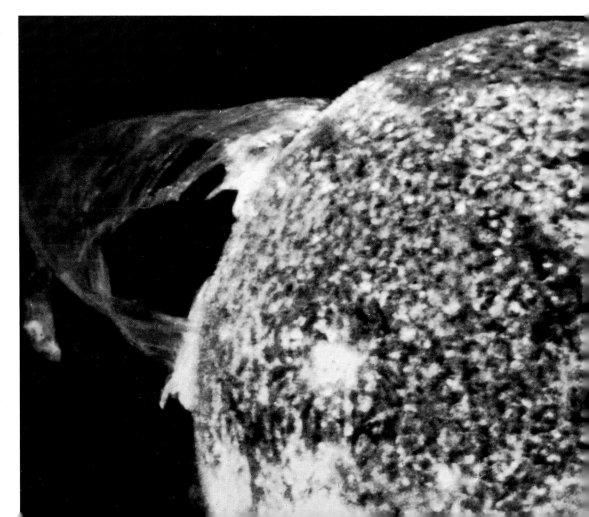

▶ This special picture of the Sun was taken in 1973 from Skylab, a space station in orbit round the Earth. It shows up the 'graininess' of the Sun's surface and a giant plume of gas erupting from a solar flare.

Sunspots

Close-up the Sun looks like a bubbling cauldron as hot gases gush out then fall back. Some of the gas streams away from the Sun into the space between the planets. The faint halo of glowing gas round the Sun is seen only in a total eclipse.

Sometimes there are dark blotches on the Sun's yellow disc. These sunspots can be many thousands of kilometres across and last for several weeks. The average number of spots on the Sun goes up and down over about eleven years. When there are a lot of spots, the Sun is active in other ways too.

Giant tongues of hot gas leap out violently from the Sun to heights of 1 million km (620,000 miles) or more. They are called prominences. Near to sunspots, flares like enormous lightning flashes can burst out. Particles shot out from the Sun take about two days to travel through space to Earth. When they reach us they cause the lights in the night sky called the aurora and can interfere with radio signals on the Earth.

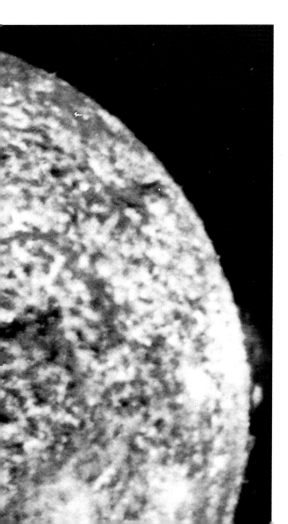

ECLIPSES

When the Earth, the Moon and the Sun line up in space, we see an eclipse. If the Moon gets between us and the Sun there is an eclipse of the Sun. If the Earth is between the Sun and the Moon there is an eclipse of the Moon. By chance, the Moon and the Sun look the same size.

▼ A total eclipse of the Sun. All you can see is the faint white haze around the Sun, called the corona.

Eclipse of the Sun

There is an eclipse of the Sun when the Moon passes directly in front of the Sun and the Moon's shadow falls on the Earth. Occasionally the Moon covers the Sun and makes a total eclipse lasting a few minutes. Partial eclipses, when the Moon covers part of the Sun, are more common than total eclipses.

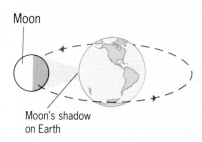

Eclipse of the Moon

There is an eclipse of the Moon when the Sun, Moon and Earth line up and the Moon passes through the Earth's shadow. Eclipses of the Moon are visible from all places on Earth where the Moon has risen. They last for several hours while the Moon goes through the Earth's shadow. During an eclipse, the Moon looks a lot darker and reddish, but it does not disappear completely.

The longest a total eclipse of the Sun could last is 7 minutes and 40 seconds, but one so long has never been observed, and they are normally much shorter.

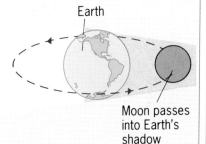

Stars

◀ The light from red and blue stars, reflecting off the gas clouds around them make this part of the sky unusually colourful. The brightest star at bottom left is Antares. Just to the right of it is a globular star cluster called M4. Above Antares you can see dark dust clouds blotting out the light from stars behind.

The brightest star in the sky is Sirius, sometimes called the 'Dog Star'. The nearest star (apart from the Sun) is Proxima Centauri, which is 4·3 light years away. It belongs to a triple star, Alpha Centauri. The other two stars are only slightly further away.

There are about 5,780 stars that can be seen by eye without a telescope. About 2,500 are visible from one place on the Earth at any one time (on a clear dark night).

On a clear, dark night you can see hundreds of twinkling stars in the sky. These stars are huge glowing balls of gas like our Sun but they are fainter because they are so much further away. The light from even the nearest stars takes years to reach us. We look up at the stars through air that is constantly blowing about, so their light is unsteady and they seem to twinkle.

A star shines because nuclear energy is given off as hydrogen gas particles crash into each other in the centre of the star. Scientists call what happens nuclear fusion.

Our Sun is one ordinary star among millions. They are all speeding through space but look still to us because they are so far away. The patterns they make in the sky stay the same. You might sometimes see what people call 'shooting stars' flash across the sky. They are not really stars at all but meteors, which are chunks of rock and bits of dust burning up as they fall through our atmosphere.

How stars are made

Stars are being created all the time. They start off as clumps of gas and dust in space. Once the material begins to collect, the force of gravity makes it pull together even more strongly. In the middle it gets warmer and denser until the gas is so hot and squashed up that nuclear fusion can start. When this happens, a new star is born.

Astronomers look for new stars in the gas clouds of space. Often, lots of stars form near to each other in a giant cloud and make a cluster. The smallest stars contain about one-tenth the amount of material in the Sun. The biggest may be fifty times more massive than the Sun.

How stars change

Stars do not live for ever. They make nuclear energy from the hydrogen gas in their cores where it is very hot, but a time comes when all the fuel is used up. When this happens the star changes and eventually it dies. Old stars swell up into red giants. They can blow off some of their gas into space, like a big smoke ring. Astronomers can see stars like this at the centres of shells of glowing gas.

The Sun is already about 5,000 million years old and is reckoned to be about half-way through its life. In the far future, the Sun will become a red giant and swallow up the planets near to it.

After that, it will shrink until all its material is squashed into a ball about the size of the Earth. It will then be a white dwarf and gradually fade away. Stars rather more massive than the Sun finish up with a tremendous explosion, called a supernova.

Supernovas

When a supernova goes off, it shines as brightly as millions of Suns put together for a few days. The inside of the exploded star that is left becomes a bleeping radio star, called a pulsar.

Nearby supernovas in our own Galaxy are rare. Only three supernovas have ever been recorded: by Chinese astronomers in 1054, by Tycho Brahe in 1572 and Kepler in 1604. Astronomers can spot them quite often in distant galaxies because they are so bright.

In 1987, a supernova exploded in a galaxy very near to our own called the Large Magellanic Cloud, which can be seen from countries in the southern hemisphere. It was very bright, so astronomers could study this important event carefully.

Double stars

The Sun is a single star on its own, but that is quite unusual among the stars. Most are in pairs. The force of gravity keeps them together and they orbit round each other like the planets going round the Sun. The brightest star in the sky, Sirius, is a double. The nearest stars to the Solar System, Proxima Centauri and its two partners, make a triple star called Alpha Centauri. There is a famous 'double double' of four stars in the Lyre.

▲ The Helix Nebula. The star at the middle of the pink ring has blown off a shell of glowing gas into space.

Star names

Many of the bright stars have their own names. Most of them have been handed down to us from Arab astronomers of centuries ago. Arabic names often start with the two letters *Al*, such as Altair, Aldebaran and Algol. Others come from Greek or Latin, such as Castor and Pollux, the 'heavenly twins' in the constellation Gemini.

Stars are also called after the constellation they are in, with a Greek letter in front. Alpha Centauri, the nearest star, is one example.

Fainter stars do not have names. Astronomers know them just as numbers in catalogues.

▲ The Pleiades, a cluster of several hundred stars. They formed out of a cloud of dust and gas about 50 million years ago. The blue haze is starlight reflected off the remaining part of the cloud.

The planets, which are much nearer the Earth, shine with a steadier light than the stars. They look more like bright flat plates.

STAR GAZING

You can start on any clear, dark evening, so long as you have a safe place near home. Try to keep out of street and house lights; then your eyes get adapted to the dark and you see more stars. See what patterns you can imagine in the stars. Notice bright ones and faint ones. See if you can find some red stars. You will find some star charts in the article: Constellations.

Through the ages, people have imagined patterns in the stars and have given them special names. Orion, shown here, was a great hunter, son of the sea-god Neptune in Greek myths.

Star colours

If you look carefully at the bright stars, you might notice that some of them look quite red while others are brilliant white or bluish; our Sun is a yellow star. The colours show up even better on photographs.

The stars shine with different colours because some are hotter than others. The Sun's surface is about 6,000°C. The red stars are cooler and the blue-white ones hotter, at about 10,000°C or more.

Giants and dwarfs

Although the stars are so far away that they look only like points of light, even in the world's biggest telescopes, astronomers have worked out that stars cover a huge range of sizes. Astronomers often call the large ones 'giants' and the small ones 'dwarfs'.

A supergiant star where the Sun is would swallow up the planets out as far as Mars. The Sun is a smallish star, though some are even smaller. The unusual kind of star called a white dwarf has a diameter less than one-hundredth that of the Sun. In contrast, there are some truly immense stars that have puffed themselves out till they are several hundred times the Sun's size. The bright red star called Betelgeuse in the constellation of Orion is about 500 times bigger than the Sun. If Betelgeuse were at the middle of the Solar System, it would swallow up the Earth and reach nearly as far as Jupiter.

Star clusters

Stars often form together in families, called clusters. One of the easiest to see is called the Pleiades or Seven Sisters and is in the constellation of Taurus, the Bull. There are six bright stars and many more come into view in a telescope. Photographs of the Pleiades show up the shining gas between the stars. There are many other beautiful clusters similar to the Pleiades, each with a few hundred stars in it.

Constellations

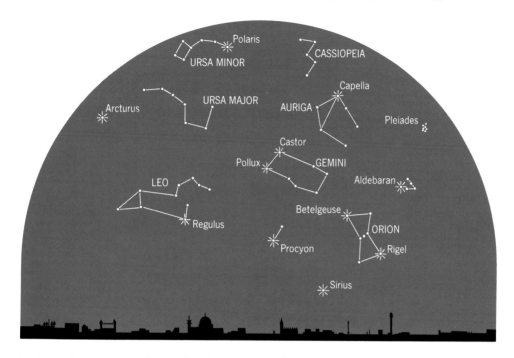

◀ Some of the constellations visible from northern countries. You would see the sky looking like this from Britain if you stood facing south around midnight in January.

Stars in a constellation appear to us on the Earth to be in a group. In reality, most are vast distances away from one another.
Largest
Hydra, the Water-snake
Smallest
The Southern Cross (Crux)
With brightest stars
Orion and the Great Bear (Ursa Major)
Dimmest
Mensa, the Table

If you look at the stars in the night sky, it is not hard to pick out patterns, such as squares, triangles and crosses, made by the bright ones. For thousands of years, people have been giving names to groups of stars that seem to make patterns. A named group of stars is called a constellation. Some names we use today were given by the Greeks at least 2,000 years ago, and the Arab and Chinese astronomers had their own constellations too.

Many of the old names are of animals, like the Bull and the Swan, and of people from Greek myths, such as Perseus and Andromeda. Among the recently named constellations you will find the Microscope and the Clock.

There are 88 constellations and they cover the whole sky. Astronomers have agreed where the imaginary boundary lines between the constellations are, and the official names they use are in Latin.

Which constellations you can see varies according to the season, the time of night and where you are on the Earth. If you live in the northern hemisphere, some of the easiest constellations to find are in the sky on winter evenings. Orion, the Hunter, has three very bright stars in a row making his belt. Near him you can find Taurus (the Bull) and Gemini (the Twins). On summer evenings, three of the brightest stars are in the Eagle, the Swan and the Lyre. They stand out as a giant triangle and can help you find some of the other constellations. From the southern hemisphere people see different constellations, such as the Southern Cross and the Centaur.

▼ Some of the constellations visible from southern countries. You would see the sky looking like this from Australia if you stood facing south-east at midnight in October.

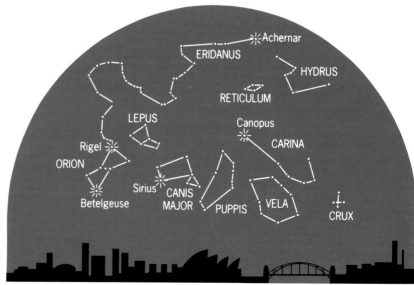

Pulsars, Nebulas, Quasars

▲ The Orion Nebula is 1600 light years away from Earth. Vast clouds of dust and gas, mainly hydrogen, are lit up by the Trapezium Cluster of hot, young stars in its bright glowing centre.

Pulsars are a kind of incredibly dense star called a neutron star. A cupful of neutron star material weighs a million million tonnes.

The fastest pulsars spin several hundred times a second, making one flash each turn.

Pulsars are strange stars that send out radio signals as a series of quick, short bleeps. They were discovered in 1967 by radio astronomers in Cambridge, England.

The time between the bleeps is different from one pulsar to another: it ranges from a few seconds to a small fraction of a second. A pulsar acts a bit like a light-house. It is a spinning star sending out two beams into space in opposite directions. Radio astronomers can catch the beams as they sweep past their telescopes.

Pulsars are the remains of supernovas, ordinary stars that reach the end of their lives and explode.

The most famous pulsar is in the middle of the Crab Nebula, where its energy lights up a patch of glowing gas in the constellation Taurus. Chinese observers saw a star explode at this point in the sky in the year 1054. The Crab pulsar sends out 30 beeps every second. As well as the radio pulse, light and X-rays from the star are switching on and off at the same speed.

Clouds of gas and dust

Astronomers use the word nebula for a thing in the sky that is not a star but looks like a misty patch of light. Our Galaxy is full of glowing clouds of gas and dust between the stars. Some give out their own light. Others reflect light from nearby stars. Some clouds are dark and black. We can see they are there because they blot out light from stars behind them.

Quasars

Quasars were first detected as very strong radio signals coming from particular places in the sky. Seen with an ordinary telescope, they were just faint points of light. At first they were called 'quasi-stellar radio sources', which means something sending out radio waves that looks like a star, but is not. Soon people called them 'quasars'.

Quasars are the most distant objects in the universe that we can see. The most distant ones are thousands of millions of light years away. The light has been travelling to us for so long that it was half-way here before even the Sun and the Earth were formed. We are seeing quasars now as they were when the universe was much younger.

Scientists think they may be the centres of galaxies where there are giant black holes. Huge amounts of energy would be given off by stars falling into a black hole (see page 60 to find out about black holes).

Comets and Meteors

◀ Halley's Comet in March 1986. The coloured dots are stars. They look coloured because the photograph is actually made up from three separate coloured pictures, one red, one green and one blue.

A bright comet in the night sky looks like a hazy patch with a long wispy tail. Comets bright enough to be seen easily by eye are quite rare, but astronomers using telescopes find 20 or 30 every year.

Comets are made of ice, gas and dust. They are out in space between the planets, travelling in orbits round the Sun, and they shine because they reflect sunlight. The long tails grow only when comets get near the Sun. They are caused by the light and streams of atomic particles from the Sun.

When a comet comes near Earth, it gets brighter for a few weeks. A check from night to night shows that it moves slowly among the stars. Then it gradually fades and disappears.

Most comets are seen only once. They are pulled towards the Sun by the force of its gravity but afterwards they vanish into distant space. Others get trapped by the gravity of the Sun and planets so that they keep going round the Sun in long oval orbits. The most famous comet of all, Halley's Comet, is like that. It can be seen from the Earth every 76 years. The last time was in 1986. The comet is named after Edmund Halley, who saw it in 1682. He was the first person to realize that it came back regularly. He used old records of observations going back to 1301. The earliest definite observation of Halley's Comet that we know about was made in China in 240 BC.

In space, between the planets, there are particles of dust and pieces of rock. When they collide with the Earth, they get red-hot as they push their way through our atmosphere. Most burn up completely and we just see the flash. These are meteors. The common name for them is 'shooting stars', but they are not really stars.

As the Earth travels round the Sun, it sometimes goes through swarms of meteors. On these days we get showers of meteors, which all seem to come from a small area of the sky.

Meteorites and craters

Some pieces of rock from space are big enough to fall to the ground. They are called meteorites. There are three main kinds: some are stony, some are nearly all iron and some are a mixture. They are different from the rocks the Earth is made of, so scientists like to study them.

A really large meteorite would make a big round crater in the ground. Meteorites made the craters on the Moon and planets. The most famous crater on the Earth is in Arizona in the United States. It is 1·2 km (¾ mile) across and was made about 25,000 years ago.

Long ago, when the Solar System first formed, many more large pieces of rock were drifting in space but we think they have all fallen onto the planets by now.

▲ One of the many pieces of iron meteorite found in the area around Meteor Crater in Arizona. This one weighs 35 kg. Most of the original meteorite burnt up when it hit the ground.

Largest known meteorite
Called the Hoba meteorite, it was found in 1920 in Hoba West in south-west Africa. It measures 2·8 m by 2·4 m and weighs about 60 tonnes.

About 500 quite large meteorites strike the Earth each year.

Galaxies

▶ The spiral galaxy in Andromeda is just over 2 million light years away and is our nearest big galaxy. There are two small elliptical galaxies close to it.

▶ An elliptical galaxy, called M49, photographed with a large astronomical telescope. This giant ball of stars is 42 million light years away and measures 50,000 light years across.

Our Sun and all the stars you can see in the night sky belong to a giant family of stars that we call our Galaxy. On a dark night, the Milky Way looks like a hazy band across the sky. It is the light from countless stars in the Galaxy.

If you could go far out in space to look at the Galaxy, you would see a flat, circular shape with a bulge in the middle and a pattern of spiral arms. Between the stars there are giant clouds of gas and dust, some of them shining brightly and some dark. Beyond our own, there are many more galaxies scattered all through the Universe.

Kinds of galaxies

Ours is a spiral galaxy, about 100,000 light years across. Many large galaxies have beautiful spiral shapes. There are also elliptical galaxies that are just flattened balls of stars. The third kind are irregular, which means they have no particular shape. Some are very strange and look as if they have exploded or might be two galaxies colliding. Many of these galaxies send out strong radio waves that can be picked up by radio telescopes. Galaxies tend to cluster together in space. Our own Galaxy belongs to a cluster that we call the Local Group.

Some well-known galaxies

The nearest galaxies to us are the two Magellanic Clouds. They look like little clouds of stars and can be seen only from the southern hemisphere. They are about 180,000 light years away. The biggest galaxy in our Local Group is called the Andromeda Galaxy. It can just be seen by eye as a misty patch. It is a spiral 130,000 light years across. One of the biggest galaxies that astronomers have studied is an elliptical one called M87. It is a million light years across and 50 million light years away.

Universe

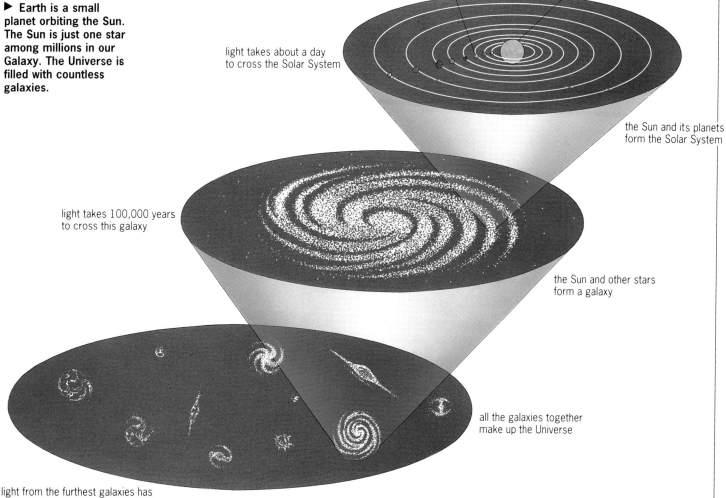

▶ Earth is a small planet orbiting the Sun. The Sun is just one star among millions in our Galaxy. The Universe is filled with countless galaxies.

We use the word 'Universe' to mean everything that exists, from the Earth out to the most distant parts of space that astronomers can possibly see. People used to think that the Earth was the centre of the Universe. Although the Earth is important to us, we know now that it is a little planet going round the Sun, which is just one of millions of ordinary stars in our Galaxy.

The Universe contains countless galaxies of stars. The furthest ones we can see are so distant that their light takes thousands of millions of years to reach us. This delay means that we are seeing them now as they actually were long ago. When we look deep into space, we look a long way back in time as well.

We can also tell that the galaxies are rushing apart and that the ones furthest away are going the fastest.

Beginning and end

Astronomers have worked out that the Universe we know started thousands of millions of years ago. It is changing and getting bigger all the time as the galaxies rush apart. Thousands of millions of years in the future, the stars may stop shining and perhaps the Universe will get smaller again.

Light travels at nearly 300,000 km (186,000 miles) every second. Yet light from the furthest known galaxies has taken more than 10,000 million years to reach us.

Big bang, Black holes

▲ Artist's impression of the massive black hole which may lie at the centre of our galaxy, the Milky Way. Gas, dust and stars are gradually being drawn into the black hole by its immense gravitational field.

A black hole with the mass of the Sun would be about 6 km (4 miles) in diameter. One with the mass of the Earth would be about 18 mm ($^3/_4$ in) across.

Most scientists believe that the Universe started between 10 and 15 thousand million years ago with an event that they call the 'big bang'. We imagine that this was some kind of gigantic explosion in which all the matter and energy in the Universe was created.

At first the Universe was incredibly dense and hot. Later, as it expanded outwards, the galaxies and the stars formed. The whole Universe has gone on expanding ever since. We do not know whether it will go on growing for ever or start to get smaller again.

Cosmic beings

Our bodies are made up of elements formed in the nuclear reactions at the centres of stars and in the giant explosions of dying stars called supernovas. These elements may later have become part of the giant clouds of dust that drift through space, until they were again brought together to form our Milky Way galaxy, the Solar System and the Earth.

Black holes

'Black hole' is the name astronomers use for a very strange kind of star. It cannot send out any light but it is so dense that it drags in anything that gets near enough with the incredibly strong pull of its gravity. Anything that falls in just disappears.

Astronomers think there are some black holes in space, but it is not easy to be sure because they cannot be seen! You have to look for material falling into a black hole. X-rays are given off when that happens, and some of the X-ray stars astronomers have found may be black holes.

Super-gravity

Black holes are probably made when a really massive star explodes. The inside of the star, left over after the explosion, falls in on itself and keeps falling until the material is squashed out of existence. When you throw a ball up in the air, it falls back because the Earth's gravity is pulling on it. When something weighing three times as much as our Sun collapses, its gravity is so strong that even light rays cannot get away. Astronomers think there may be black holes millions of times more massive than the Sun at the centres of some galaxies.

Observatories

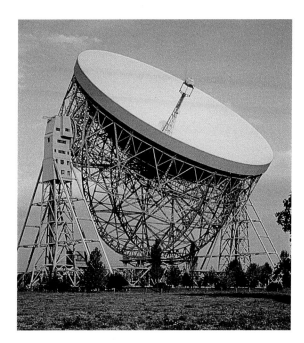

Observatories are the special places where astronomers make their observations of the sky. People all over the world have been building observatories for thousands of years, even before telescopes had been invented. They were places for following movements and changes in the sky. A modern observatory is usually a building containing a telescope, or a group of buildings with different telescopes, offices, workshops and bedrooms. But there are also unmanned observatories orbiting the Earth on satellites. In the United States, there is even an observatory in a high-flying aircraft.

There are two main kinds of ground-based observatories. Optical observatories have ordinary telescopes for looking at the light from the stars. At radio astronomy observatories there are big dishes to collect radio signals from space.

Radio observatories are very different from optical observatories. Radio telescopes do not need to be inside a building. The radio waves from space come through the clouds so it is not very important where you build the observatory. The telescopes can be used most of the time, in daylight as well as at night. One of the best-known radio observatories is at Jodrell Bank in England. The world's most powerful radio telescope, in New Mexico in the United States, has 27 dishes.

In contrast, optical observatories need to be built in places a long way from the street lights of cities and where there are not many cloudy nights. The air must be very clear and still so that the stars do not twinkle too much and the telescopes can see very faint things. The best sites are on mountain-tops in warm places. There are big observatories at Kitt Peak in the Arizona Desert in the United States, on top of volcanoes in Hawaii and the Canary Islands, and in the Andes Mountains of South America.

The telescopes are housed in big domes. At night, a slit in the dome is opened and the whole dome can turn so that the telescope can point to different parts of the sky. The newest telescopes are run by computers and the astronomers can work from a warm control room instead of in the cold open dome.

◀ The Lovell Telescope at Jodrell Bank in Cheshire, England, has a massive dish 76 m across to catch radio waves from space. The radio signals are focused onto the receiver on the tower in the middle.

▼ A large astronomical telescope inside a dome. This one is the Anglo-Australian Telescope in New South Wales, Australia. It is a reflecting telescope and its main mirror is 3.9m across.

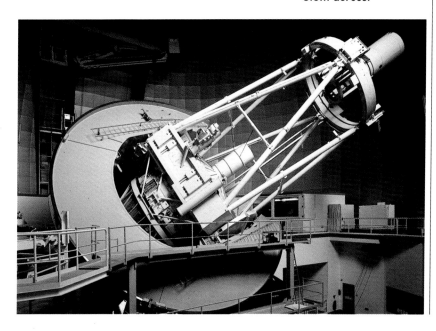

Energy

Nothing can live, move or work without energy. Plants need it to grow. We need it just to keep alive. Our energy comes from the food we eat. Machines need a supply of energy to work. Some use electricity, while others burn fuels which contain stored energy. We all use energy and almost all of it originally came from the Sun.

Different types of energy

There are many different kinds of energy. Heat, light and sound are all kinds of energy which move around. Light and heat are radiant energy, often called radiation. Other kinds of radiation include microwaves and X-rays. Every time we turn on an electric switch we use electrical energy. It provides lighting and heating and makes many things work. Chemical energy is stored in foods and fuels. It is released when fuels are burned, or chemicals get to work on the food in our bodies.

When some atoms split or join to make different atoms, huge amounts of energy may be released. This is called nuclear energy. We can use it peacefully to make electricity, or destructively in bombs.

Anything that moves has kinetic energy. The larger the moving thing and the faster it moves, the more kinetic energy it has. If you throw a ball up in the air, you give it kinetic energy. As it goes up, it slows down and loses kinetic energy. But it gains a kind of stored energy called potential energy. When the ball stops at the top of the throw, it has no kinetic energy because it is not moving; but it has potential energy. As it falls down again, its potential energy is changed back into kinetic energy.

Keeping energy

Sometimes energy seems to be used up. But it is never lost. It just changes into another kind of energy. A ball rolling along the flat ground slows down and stops, losing its kinetic energy. It slows because it rubs against the ground; we say there is friction between the ball and the ground. The friction heats up the ball and the ground. The ball's kinetic energy has changed into heat, but the total amount of energy has not changed.

Storing energy

Lots of things store energy, including food. Our food is either plants or animals which have been fed on plants. When plants grow, they take in and store energy from the Sun. The Sun's energy has also been stored in the fossil fuels: coal, oil and gas. Millions of years ago these were living trees

Measuring energy
We measure energy in joules. Power, the amount of energy used every second, is measured in watts.

Energy for people
We measure food energy in kilojoules (kJ). One kilojoule equals 1,000 joules. An average 11-year-old needs about 10,000 kilojoules of energy every day.

Energy for action
A child weighing 30 kg uses about 700 joules of energy walking upstairs.

Using electrical energy
An ordinary electric light bulb uses 100 joules of energy every second (100 watts). A colour TV uses about 200 joules every second (200 watts).

Energy for living
The food we eat gives us energy. We need it, even when we are asleep, to stay warm and keep the organs of our body working. When we are active, we need more energy to move our muscles.

Energy in plants and animals
Green plants use the Sun's energy to make their food from simple materials like water and carbon dioxide gas in the air. Like other living things, plants need food to live and grow. We get our energy by eating plants, or eating meat from animals which have fed on plants. So our energy really comes from the Sun.

Energy from the Sun

Almost all our energy originally came from the Sun. At the Sun's centre, nuclear reactions give enormous amounts of energy which radiate from the surface as heat and light. There is enough nuclear energy left in the Sun to keep it shining for another 5,000 million years.

Energy from the Sun
On average, each square metre of the Earth's surface receives the same energy from the Sun as it would do from a one-bar electric fire.

and tiny sea creatures storing energy from the Sun as they grew. Once fossil fuels have been taken from the ground they are never replaced, so we must use them carefully and not waste them.

Bottled gas is a useful way of storing energy. So is a battery. When a battery is properly connected, chemical changes inside it produce electricity. A dam across a river is another way of storing energy. The water which collects in a lake behind the dam has potential energy. When electricity is needed, some of the water trapped by the dam flows down a channel and turns a turbine to make electricity. The stored potential energy becomes kinetic energy in the flowing water before changing to electrical energy.

Energy in fossil fuels
Millions of years ago, plants and tiny creatures took in energy from the Sun as they grew. When they died, they became buried under layers of rock which slowly formed above them, and were gradually turned into coal, oil and natural gas. These fossil fuels now store energy which once came from the Sun. One gram of coal contains about 25 kilojoules of energy.

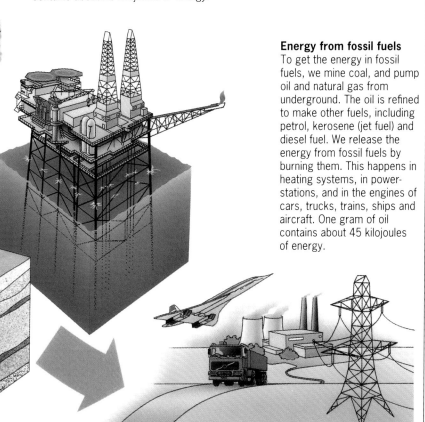

Energy from fossil fuels
To get the energy in fossil fuels, we mine coal, and pump oil and natural gas from underground. The oil is refined to make other fuels, including petrol, kerosene (jet fuel) and diesel fuel. We release the energy from fossil fuels by burning them. This happens in heating systems, in power-stations, and in the engines of cars, trucks, trains, ships and aircraft. One gram of oil contains about 45 kilojoules of energy.

ENERGY AND THE HOME

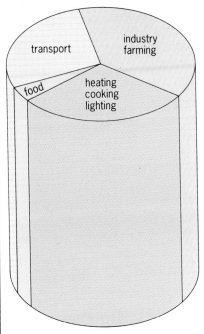

How energy is used for each person in an industrialized country.

▲ Compare the size of the pie charts. Each person in an industrialized country uses over ten times as much energy as someone in a developing, Third World country.

▲ *Solar Challenger* has special panels on its wings to absorb the energy in sunlight and make electricity for its motor. But all aircraft use energy which originally came from the Sun.

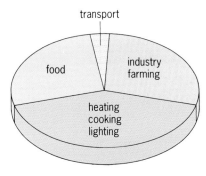

How energy is used for each person in a developing, Third World country.

How much energy do we use?

Everyone needs energy just to keep alive. However, we also use a lot of energy running machines to help us in our work and play. Almost all homes in Europe, North America, Australia and other industrialized countries have a supply of electricity for lighting and heating and to run the TVs, washing machines and other electrical machines. Pipes usually bring a supply of gas as well for cooking or heating. We also have cars which get their energy from petrol. In many developing countries people are much poorer and use a lot less energy.

Alternative energy

Most of the energy we use today comes from the fossil fuels: coal, oil and gas. But these will not last long because they are not being replaced. Also, burning them is slowly harming the atmosphere. Engineers are now looking for other ways of supplying energy. Modern windmills are being built in groups to produce electricity from the wind. In some places, the sea water flowing to and fro with the tides will also turn turbines, and even waves on the sea can produce electricity. The Sun's energy can be collected by solar panels which heat water, or by solar cells which produce electricity when light falls on them. Geothermal energy comes from the hot rocks inside the Earth which can heat water and make electricity. All these methods can supply our energy and they do not harm the atmosphere. But people go on using fossil fuels because they are still the cheapest and most convenient way to get energy.

Flashback

Thousands of years ago people had only the Sun's energy and their own energy. They burnt wood for heat, and animals provided energy to carry things and work on the farms. Then they learnt to use energy in rivers to turn water wheels, and the energy in wind to drive windmills and sailing ships. About 200 years ago they began to burn fossil fuels.

Wind power

The wind can be very destructive. Gales can uproot trees and lift tiles off roofs. But the wind can also be put to work. Sailing ships and yachts have sailed round the world on wind power alone. And windmills have used the power of the wind for grinding corn and pumping water. Today, aerogenerators are using the wind's energy to generate electricity. Unlike oil and gas, the wind is one source of energy which will never run out.

Aerogenerators

Most aerogenerators have a tall, slim tower with huge blades like an aircraft's propeller mounted on top. The blades can be over 20 m (65 ft) long. As they spin in the wind, they turn a generator which produces electricity. Aerogenerators are placed on exposed, windy sites often in large groups called wind farms. Unlike fuel-burning power-stations, they do not pollute the atmosphere, but the force of the wind is unpredictable and few sites in the world are suitable.

Windmills

A hundred and fifty years ago, there were thousands of windmills across The Netherlands and the fens of eastern England. Many were used to drain water from low-lying land. Very few remain today, but some have been kept in working condition. They have vanes (sails) which turn a shaft. This is connected by gearwheels and another shaft to a pump, a stone for grinding corn, or some other machinery. When the wind changes direction, the sails must be turned so that they face into the wind. In post mills, the whole building turns on a vertical post. In later designs, called tower mills, only the top part of the tower turns. The main tower, made of wood or bricks, is fixed.

Windmills around the world

Windmills are still used in many parts of the world for pumping water from wells and drainage ditches. Some have fabric or slatted wooden sails. Others are made by cutting an old oil drum in half and attaching the two parts to a vertical shaft. In Australia and America, one common type of windmill has a tall, framework tower. The rotating part has lots of steel blades with a tail vane to turn it into the wind.

The earliest known windmills were used for grinding corn in Iran in the 7th century AD.

In 1840, there were around 10,000 working windmills in England and 8,000 in Holland.

▼ A wind farm at Altamont Pass, California, USA. This farm has 300 wind generators contributing electricity to the American national grid.

Solar power

Light and heat from the Sun pour down on the Earth all the time. When we turn this energy into electricity or use it as heat, we call it solar power. On a sunny day, a square patch of Earth facing the Sun with sides 1 m (40 in) long gets up to 1,000 watts of power from the Sun: enough to run one bar of an electric fire. In fact, the Sun could supply all the power we need for the whole world if we could collect it and use it efficiently. The equipment needed to turn the Sun's energy into useful power is expensive but it costs less to run and maintain than an ordinary power station.

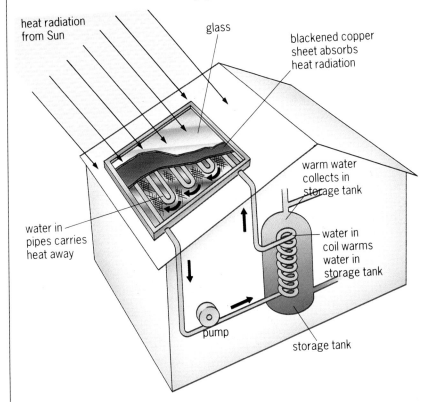

◀ A solar panel absorbs heat radiation from the Sun. The heat helps to warm water for the house. There is less for the main heating system to do, so fuel bills are reduced.

Electricity from sunlight

Electricity is probably the most convenient type of power we use every day, and solar cells can turn sunlight directly into electricity. Solar cells are made from thin slices of pure silicon, a material which can be got from sand. The top of the slice is a slightly different kind of silicon from the bottom, and when light shines on it, an electric current will flow along a wire connecting the top to the bottom. A single solar cell produces only a tiny current, but an array of cells connected together makes a useful amount of power. Satellites in space use huge panels of solar cells to supply their electricity. In remote parts of some developing countries, solar cells provide electricity to pump water for drinking and growing crops and to power refrigerators storing medicines.

Using the Sun's heat

If you have ever sat in a car on a hot sunny day you will know that the Sun's energy can be trapped as heat. Solar panels on the roofs of buildings can also trap this heat to give us hot water. In the solar panel, under a sheet of glass, are pipes fixed to a black plate. The Sun heats up a liquid in the pipes and this liquid heats up a tank of water.

Huge solar furnaces use the Sun's heat to make electricity. A field of mirrors collects sunlight and concentrates it onto a furnace where the heat boils water to make steam. This drives a turbine making electricity just as in an ordinary power station.

Flashback

Energy from the Sun has always been important to people. Over 2,000 years ago the Greeks and Romans were building their houses to face the Sun. In 1714 Antoine Lavoisier, a French scientist, made a solar furnace which could melt metals. The first steam-engine to work on solar power ran a printing press in Paris in 1880. By 1900 many houses in the hotter parts of the USA had solar water-heaters. All these inventions used the heat from the Sun. It was not until 1954 that the first practical solar cells turned sunlight directly into electricity.

Geothermal power

The centre of the Earth is very hot. The Earth is a huge natural store of heat. In some places the temperature of the rocks in the Earth's crust may increase by 55°C per kilometre or more. In a few places the heat is close enough to the surface for people to use this 'geothermal' power.

Water which soaks down into the ground may be heated and then come to the surface as steam from cracks in the ground, or as geysers (hot springs). Such natural hot water is used for heating buildings in parts of New Zealand and Iceland.

A geothermal power station uses steam from a borehole drilled a kilometre or more down into the hot rocks. If water is not there naturally then it is pumped down from the surface, through another borehole, to be heated by the rocks. The borehole takes the place of the boiler in a fuel-burning power station. The steam drives turbines (special water wheels) which turn a generator to make electricity. Unlike most other types of power-station, geothermal stations need no fuel and do not pollute the atmosphere.

Clean but costly

Geothermal power stations are very expensive to build, about five times the cost of a nuclear power station. Engineers often have to drill several kilometres down through the Earth's crust. Here, temperatures can reach 200-300°C or more, hot enough to roast the drill bits.

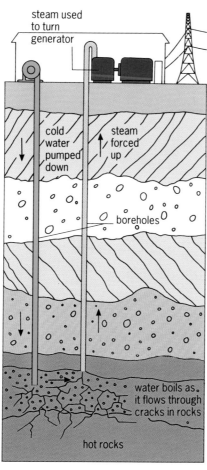

◄ Geothermal power station. Cold water pumped down one borehole is heated by underground rocks. Steam coming up the other borehole is used to turn a generator.

Geothermal comes from two Greek words: *ge* meaning 'earth' and *therme* meaning 'heat'.

◄ View over the Wairakei geothermal extraction plant showing pipes carrying steam to the power station.

Water power

Water always flows from a higher point to a lower point. This movement of water can be used as a source of energy. It can be the gentle flow of a river, or water falling from a great height as in a waterfall or from a dam. The never-ending movement of waves at sea and tides can also be harnessed to provide energy. Unlike many other sources of energy, water does not get used up and there will always be a cheap and constant supply of moving water on the earth.

Hydroelectric power

Electricity is generated when water drives a machine called a turbine which is connected to a dynamo. Turbines are more efficient versions of earlier water wheels. They are designed to take as much energy from the moving water as possible. Hydroelectric power-stations are often built in hilly regions where there is lots of rain. A lake or reservoir provides a store of water high above the generating station. The amount of power available depends on the height the water falls. A dam is often needed to increase the size of a natural lake. Water flows from the reservoir down to the turbines through strong steel pipes or tunnels.

Some hydroelectric power stations are on river systems and are built inside or at the bottom of a specially constructed dam.

Tidal power

Tides provide another source of moving water that can produce power. A dam is built across the mouth of a river in a place where the height between low and high tide is great. Water rushes through tunnels in the dam as the tide rises and flows out of them when the tide turns. Turbines are turned by this flow and electricity generated.

◀ The Hoover Dam, Colorado River, USA, is over 220 m high and 379 m long. The reservoir behind it, known as Lake Mead, is 185 km (115 miles) long, and 180 m deep. It has a hydroelectric power station at its base which supplies most of the electricity for the states of Arizona, Nevada and southern California.

ENERGY AND THE HOME

Unfortunately, high tide comes at different times each day and providing electricity when it is most needed is difficult.

Scientists are exploring ways of using the up-and-down movement of waves in the open sea as a reliable, cheap source of power.

Flashback

The earliest water wheels were invented by the Greeks around 100 BC. They were placed horizontally in the water flow. The Romans used vertical water wheels to provide the power to mill grain. There were two common types of vertical wheels. An overshot wheel turned when water fell onto it from above. In contrast, an undershot wheel was placed in a fast-flowing stream to turn it. During the Industrial Revolution large water wheels provided power for factories.

▲ A water wheel near Murcia in south-east Spain. Water wheels are one of the oldest forms of water power, believed to have been developed around 100 BC.

PUMPED STORAGE

There is a further important application of hydroelectricity: pumped storage. Electricity itself cannot be stored in large quantities. It has to be generated just as it is wanted. In a pumped storage system, pipes run between two reservoirs at different levels. When water is allowed to run down from the higher level to the lower, it passes through a turbine and generates electricity. When the demand for electricity is low, during the night, the cost of electricity falls. The turbine becomes a pump to send the water back up to the higher level, powered by the cheap electricity. Then the water can be used again to generate more electricity when the demand is greater the next day.

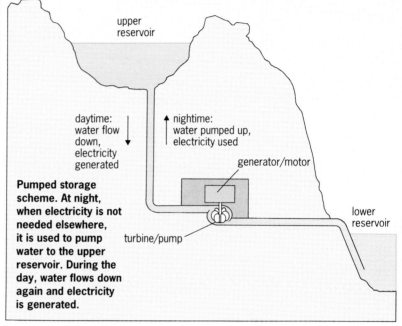

Pumped storage scheme. At night, when electricity is not needed elsewhere, it is used to pump water to the upper reservoir. During the day, water flows down again and electricity is generated.

A dam is a barrier built across a stream, river or estuary. It is used to prevent flooding, to generate electricity by hydroelectric power or to hold water for drinking or irrigating land. The dam blocks the water flow and the water collects behind the dam to form a reservoir. Modern dams are mainly of two kinds, embankment dams and masonry dams. Masonry dams are usually built of concrete.

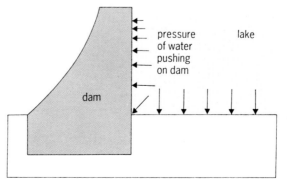

◀ A gravity dam is one kind of masonry dam. The base of the dam, where the water pressure is greatest, is made of concrete.

Nuclear power

There are over 400 power-producing nuclear reactors in operation worldwide. More than half of them are pressurized water reactors.

The world's first nuclear reactor providing electricity began working in 1956 at Calder Hall, Cumbria, in the UK.

One tonne of uranium fuel can produce as much energy as 25,000 tonnes of coal.

Percentage of electricity produced by nuclear power: 65% France
22% UK
16% USA

Nuclear power provides us with electricity. It uses the energy stored in the nucleus in the centre of atoms. In some very heavy atoms the nucleus can be split into two smaller parts. This process of nuclear fission releases an enormous amount of heat, which is used in a nuclear power-station to make water boil; the steam pushes round a machine called a turbine. This turns a generator which produces electricity.

Fission

Uranium is a metal which provides the fuel for most nuclear power-stations. A special type of uranium is used, called U-235 (it has 235 neutrons in its nucleus). When a neutron hits this nucleus it splits into two smaller nuclei. This process of nuclear fission produces invisible radiation, called gamma rays, and two or three high-speed neutrons shoot out. These neutrons can break up other U-235 atoms, releasing more neutrons and more energy, and so the process continues. This is called a chain reaction.

Reactors

The process of nuclear fission is very dangerous. So much energy is produced that there can be an explosion, and this is what happens in an atom bomb. In a nuclear power-station, fission is controlled so that energy is produced without

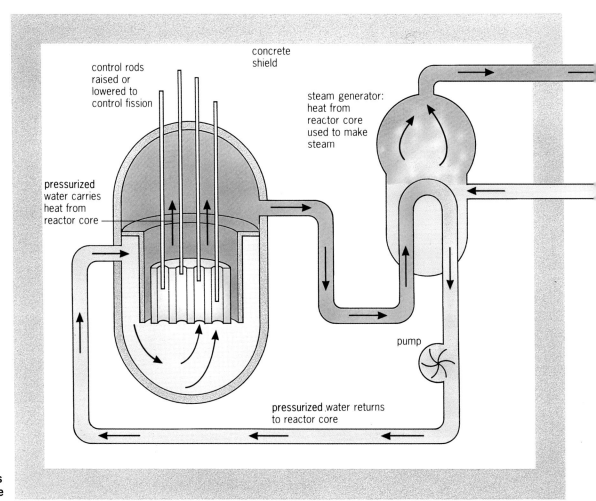

▶ **Pressurized water reactor (PWR). Heat from the reactor core is used to make steam. The steam turns the turbines which drive the generator.**

explosions. The uranium fuel forms the 'core' of a nuclear reactor. Special 'control rods' can be raised or lowered into the uranium. These rods, made of cadmium or boron, absorb neutrons. This slows the reaction.

Nuclear reactors get very hot. There are many different ways of taking the heat away to produce the steam which drives the turbines. In a pressurized water reactor (PWR), water is used at high pressure. In an advanced gas-cooled reactor (AGR), the gas carbon dioxide is passed over the fuel in the reactor core.

Some nuclear reactors can actually turn ordinary uranium into more nuclear fuel. They are called 'fast breeders'.

Nuclear waste

The fuel in a nuclear reactor is very radioactive. It produces a lot of dangerous radiation which is extremely harmful to all living things. Some of the radioactive substances produced reactor remain dangerous for thousands of years. Getting rid of this dangerous nuclear waste safely is a serious problem.

Safety and risks

A nuclear reactor cannot explode like an atom bomb but an accident at a nuclear power-station can have disastrous effects over a large area. In 1986 a major accident at Chernobyl nuclear power station in the USSR released radioactive substances into the atmosphere. Winds carried them across Europe and rains washed them down to Earth. Grass and crops became radioactive and sheep grazing on this grass were not safe to eat. There was an increase in the amount of radiation surrounding us. Even so, the risk to our health is tiny: about the same as smoking one-eighth of a cigarette a year. However, people living near Chernobyl have suffered much more, and some have died.

Nuclear power could provide electricity for hundreds of thousands of years, but is it safe? Should we build more nuclear power stations and increase the chances of terrible accidents? The problem is that, without nuclear power, it will be difficult to provide enough electricity for all future needs. Some people say that we have to accept the risks. Others say that the risks are too great; we should find other ways of generating electricity or adopt a simpler life-style that requires less electricity.

A typical nuclear power station produces about 60 tonnes of waste every year. Of this about 1 tonne is highly radioactive.

Britain has to find safe storage for over 1000 cubic metres of highly radioactive liquid waste. This amount would fill a large swimming pool.

Plutonium, formed in nuclear reactors, is the most dangerous substance known. One millionth of a gram breathed in or swallowed is enough to cause cancer.

Nuclear fusion could be a new source of electricity in the future. When the nuclei of hydrogen atoms join together, vast amounts of energy are released. This is how the Sun produces energy.

Temperatures of about 100 million degrees Celsius are needed to produce nuclear fusion.

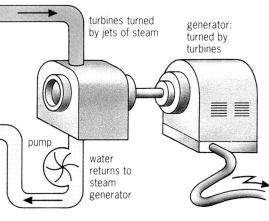

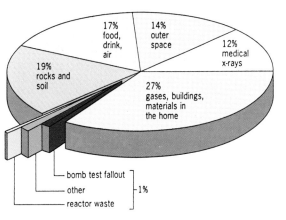

◄ There are tiny amounts of radiation round us all the time, most of it from natural sources. Scientists call it background radiation. This chart shows where it comes from.

Electricity

▶ Lightning over the town of Tamworth in New South Wales, Australia.

Typical voltages
torch battery
1½ volts

car battery
12 volts

mains power-point (UK)
240 volts

underground train
600 volts

power-station generator
25,000 volts

overhead power line
400,000 volts

lightning flash
100 million volts

▼ A battery has two terminals. Electrons flow out only if these terminals are joined by a conductor, such as a light bulb. The battery pushes the electrons from the − (negative) terminal round to the + (positive) terminal. The bulb lights up as the electrons flow through.

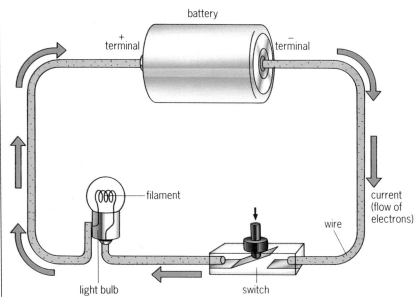

It makes cling-film stick to your hands and it crackles when you comb your hair. It can power TVs, tumble-driers and trains. It can even light up the sky in a flash.

What is electricity?

Electricity is the energy of lots of tiny particles called electrons. Electrons are parts of atoms. Everything is made of atoms, so there is electricity in everything. However, you do not notice its effects until something makes the electrons move from their atoms.

Rubbing makes electrons move. When cling-film rubs on your hand, it pulls electrons away from atoms in your skin. The atoms try to pull the electrons back again, so the cling-film sticks to your hand. The same thing happens when you rub a balloon on your sleeve. The balloon will stick to your body or to a wall.

Batteries make electrons move. They can push electrons through wires made of copper and other metals. Materials like copper which let electrons flow through are called conductors. The tiny electrons squeeze between the atoms in the wire. Some materials stop electrons passing through. These are called insulators. Plastics and rubber are insulators. Air is an insulator most of the time. But if electrons are pushed hard enough, they can even jump through air. Then you see sparks.

Circuits

When a battery is connected to a conductor such as a light bulb, electrons flow. The complete loop through the wires, bulb and battery is called a circuit. The flow of electrons is called a current. You can put an on-and-off switch in the circuit. Switching off pulls two contacts apart. This breaks the circuit and stops the flow of electrons.

Using electricity

Heat, light, magnetism and movement are just some of the things which we can get from electricity. Wires warm up when a current flows through. Thin wires warm up much more than thick ones, and some types of wire warm up more than others. In an electric fire, wire made of an alloy called nichrome becomes red-hot when a current flows through. In a light bulb, a thin tungsten wire (a filament) gets so hot that it glows white.

Every current has a magnetic field around it. If a current flows through a coil of wire, the coil behaves just like a magnet. Electro-magnets are a special type of magnet which you can switch on and off. They have a coil of wire in them. Electric motors also have a coil in them. It rotates between the ends of a bent magnet. When a current flows through the coil it becomes magnetized. The pull of the magnet makes the coil spin round. This principle is used in power-stations to generate electricity.

VOLTS, AMPS AND WATTS

Batteries and generators all push out electrons. The higher their voltage, the harder they push. Scientists measure voltage in volts.

A current is a flow of electrons. The stronger the current, the more electrons flow round the circuit every second. Scientists measure current in amperes ('amps').

Electrons carry energy. The energy comes from a battery or generator. It is spent when the electrons pass through a lamp, a kettle or some other electrical appliance.

The power of an appliance is measured in watts. The higher the power, the more energy the appliance takes every second.

There is a connection between power, voltage and current:

Power in watts = voltage x current.
 (volts)   (amperes)

So a 200 volt kettle taking a current of 10 amperes has a power of 2,000 watts (2 kilowatts).

◢ Electricians use special symbols for drawing circuits. Here are some of them:

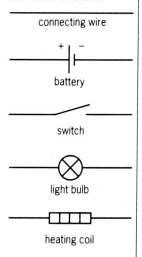

connecting wire

battery

switch

light bulb

heating coil

◂ In many power stations, the generators are turned by steam turbines. The generators produce alternating current (a.c.).

Generating electricity

Mains electricity comes from huge generators in power stations, turned by turbines driven round by the force of water or high-pressure steam. The steam is made using the heat from burning coal, oil or gas or from a nuclear reactor.

Each generator has a ring of copper coils round the outside. Inside this, huge magnets are turned by a shaft from a turbine. As the magnets rotate, they make electrons move in the coils. The current flows backwards and forwards, as first one end of a magnet and then another sweeps past. Current which flows like this is called alternating current (a.c.). The generators are known as alternators. Current from a battery is different; it flows only one way. It is known as direct current (d.c.).

Direct current (d.c.)
Current which always flows the same way. Electricity from a battery is like this.

Alternating current (a.c.)
Current which flows backwards, forwards, backwards, forwards . . . and so on. Mains electricity is like this.

Mains frequency
Measured in hertz (Hz). Tells you how many times the alternating current from the mains flows backwards and forwards every second:
UK   50 Hz
USA   60 Hz

Electricity supply

Scientists are investigating new non-polluting ways of generating electricity, including harnessing the power of tides and winds.

Large power stations can generate a 1,000 MW (1 MW = 1 megawatt or one million watts).

Batteries cannot store enough energy to keep cookers, washing machines and electric fires supplied with electricity. When you plug in a kettle, the power can come from a generator miles away. At the power-station, electricity from the generator passes along thick cables to a transformer. Here, the voltage is stepped up before the power is fed to the transmission lines which carry it across country. Increasing the voltage cuts down the current. The electrons are pushed harder, so fewer are needed to carry the power. Most transmission lines hang from tall pylons, but in areas of great natural beauty, they are buried. The lines are part of a network of power stations and cables called the Grid. If one area needs more power, it can be supplied by generators in other areas.

Distributing the power

Power from the transmission lines goes to the substations. Here, the voltage is stepped down by transformers; the lines divide to go to smaller substations ... and so on. You may have a small substation in your street. If so, it will have a steel fence

▲ Power is carried across country by overhead transmission lines. The voltage is stepped down in stages as the power is distributed to towns and factories.

Danger!

Mains electricity is dangerous. Touching 'live' (high voltage) wires and transmission lines can cause shocks, burns and even death.

ENERGY AND THE HOME

round it and a notice saying something like 'Electricity Board: keep out'. The final connection to your house is usually a cable buried under the road; in country areas, poles and overhead wires are sometimes used.

Where the cable comes into your house, you will find three things: a main switch, a main fuse which automatically switches off the current if the circuits become overloaded, and a meter so that the Electricity Board knows how much to charge you.

Power around the house

The wiring in the house divides into several branches or circuits. The wires divide at a box called the consumer unit ('fuse-box'). Here, each circuit has its own fuse or circuit-breaker.

In many houses in Britain, the power points are connected to a single cable which runs round the house and back to the consumer unit again. The cable is called a ring main. Each item you plug into it has a fuse in its plug. It is important that a fuse of the correct value is fitted. If the fuse value is too high, the fuse may not 'blow' if a fault develops.

Light bulbs in your home use between 40 and 150 watts. Fluorescent lights, although very bright may use only 25 watts.

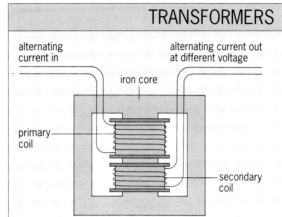

TRANSFORMERS

How a transformer works. When alternating current is passed through the primary coil, it sets up a changing magnetic field in the iron core. This changing field generates a voltage in the secondary coil. The more turns there are on the secondary, the higher the voltage produced.

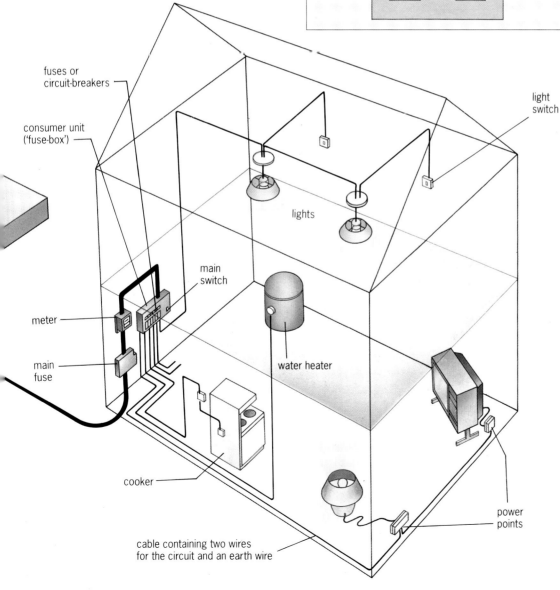

Mains cables
Each cable has three insulated wires in it. In many countries, standard colours are used in cables connected to plugs.
Live wire (brown) is at mains voltage. Neutral wire (blue) completes the circuit.
Earth wire (green and yellow stripes) is a safety wire. It stops the metal becoming 'live' if a wire works loose.

Electricity Board engineers watch TV
When popular TV programmes end, extra generators have to be ready. The demand for power can rise by 10 per cent or more as people start to make hot drinks.

◀ The main cable branches out into several cables at the consumer unit ('fuse-box'). In many houses in Britain, the power points are all connected to the same cable, called a ring main.

Batteries

The word battery really means two or more primary or secondary cells joined together. However, single cells are often called batteries; for example, a torch battery.

In a car battery, there are six separate cells all joined together in the same case.

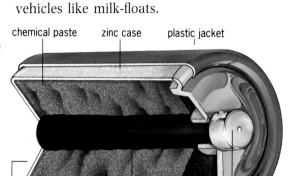

▶ A 'dry' zinc–carbon cell.

A unit in which electricity flows between a positive terminal and a negative terminal is called a cell.

Small electric batteries are used to run torches, radios, watches and pocket calculators. Large electric batteries are used to start car engines and can even drive vehicles like milk-floats.

Small batteries are usually of a kind called primary cells. The electricity comes from chemicals sealed inside them. They can be used continuously only for a few days before the chemicals are used up. Switching torches and radios off when you do not need them makes the batteries last longer. But once a battery is 'flat' and can produce no more electricity, it has to be thrown away.

Large batteries are usually secondary cells. These also use chemicals to store electricity. But when the chemicals begin to run out, they can be renewed by feeding electricity into the battery. This is called charging. It is done by connecting the battery to a dynamo (as in a car) or a charger which plugs into the mains. Some small batteries are marked 'rechargeable'. They can be charged, but only with a specially made charger. This must *not* be used on ordinary batteries.

The lead accumulator

Car batteries are secondary cells. However, instead of solid chemicals they use a liquid to carry the current from one terminal to the other. Liquids which let an electric current flow through them are called electrolytes. Some modern batteries use a jelly instead of liquid acid. This is less likely to leak.

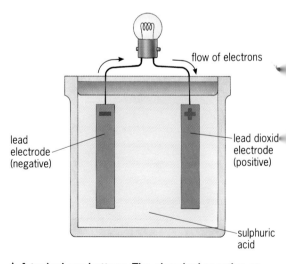

▲ A typical car battery. The chemical reaction at the lead plate gives a supply of electrons which flow through the circuit to the lead dioxide plate.

TORCHES

People have always found it convenient to be able to carry light around with them. The Romans twisted fibrous hemp or flax into a brand that could easily be held. Dipping it in oil or fat made it burn brightly. Oil lamps and lanterns replaced this simple torch, and a modern torch is powered by electricity, using electrical cells put together in a battery. Some torches are rechargeable. Some are designed to flash, or give coloured light as a warning. Some have small fluorescent tubes, giving longer battery life.

▼ A torch powered by batteries.

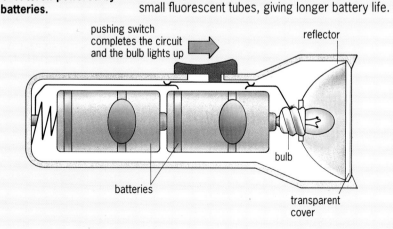

Magnets and dynamos

There are probably lots of magnets in your house. They hold refrigerator and wardrobe doors shut and they attract can-tops to the can-opener. There are other magnets you cannot see: in the door bell, the telephone, the television and the electric motors in food mixers and drills.

Magnetic materials

Magnets attract some metals, but not all. They attract iron, nickel, cobalt and most types of steel. But there are many metals they do not attract, including copper aluminium, brass, gold, tin, silver and lead.

Materials which are attracted to magnets can be *made* into magnets. If you put a steel needle next to a magnet it, too, becomes a magnet, and it stays magnetized when you take the magnet away. An iron nail also becomes magnetized near a magnet. But it quickly loses its magnetism when the magnet is removed. Steel keeps its magnetism, but iron does not. Magnets which keep their magnetism are called permanent magnets.

Magnetic poles

The forces from a magnet seem to come from two points near its ends. These points are called the poles of the magnet. One is a north pole; the other is a south pole. If you hold magnets with their north poles close they push each other apart: they repel. The same happens with two south poles. But a north pole and a south pole attract. Poles of the same kind repel; poles of opposite kinds attract. The area over which a magnet can attract or repel is called its magnetic field.

When a piece of iron or steel enters the magnetic field of a magnet, it becomes magnetized with poles that are opposite to those of the permanent magnet. This is why they are attracted to each other.

Nobody knows for certain how magnets work. Scientists think that, in materials like iron and steel, each atom is a tiny magnet. Normally, the atoms point in all directions and their magnetic effects cancel. But when a material is magnetized, its atoms line up in the same direction and it becomes one big magnet.

A giant magnet

The Earth itself acts as a magnet because it has a lot of iron in its core. If a magnet is hung up by its middle, it tries to point north-south because its poles are pulled by the Earth's magnetic poles. A magnet's 'north pole' really means 'north-seeking pole'. This is the end that tries to point north.

Electromagnets

An electromagnet is a type of magnet which can be switched on and off. It is made by winding specially insulated wire round an iron core. An electric current is then passed through the wire, which magnetizes the iron. The iron loses its magnetism when the current is switched off.

Tin is not attracted to magnets. Magnets attract 'tin' cans because these are mainly steel. The tin is just a thin coating.

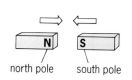

▲ When the ends of two magnets are brought close, poles of the same kind repel but poles of opposite kinds attract.

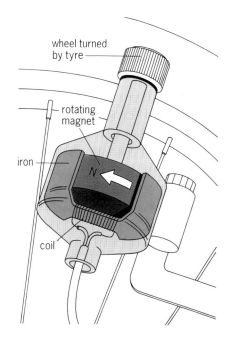

◀ Dynamos work in the opposite way to electromagnets. They produce electricity when turned, like small generators. This dynamo is turned by a bicycle tyre. Inside the dynamo, a magnet is rotated close to a coil of wire. As the magnetic field sweeps past the coil it generates electricity.

Household equipment

Two hundred years ago, life in a small house in Europe or North America was much less convenient than it is today. Water might have come from a well shared with several other families. The toilet, outside the house, was probably a wooden seat with a hole in it over a pit. Food was cooked over an open fire, and candles provided lighting. If you took a bath, it would have been in a tub on the kitchen floor. Household chores like washing and cleaning were all done by hand and could take women most of the day.

Most modern houses are connected to mains electricity, drains, water, and possibly a gas supply as well. These are called services. Certain basic equipment is built into the house, like taps, sinks, bath, toilet, and a central heating system to provide warmth and hot water. Most other equipment is not built in, but some things are so big and heavy that they are kept in one place. These include cookers, washing machines, tumble driers and refrigerators.

Refrigerators

Refrigerators are machines for keeping things cold. They move heat, taking it away from the inside, which becomes colder, and giving it to the outside, which gets warmer.

Refrigerators work by evaporation. When a liquid changes to a vapour it takes heat from its surroundings. If you wet your finger and then wave your hand, that finger feels coldest. The water takes heat from your finger as it evaporates. In a refrigerator the cooling is done by a special liquid called a refrigerant.

In a compression refrigerator the refrigerant circulates around a sealed system of pipes. An electric pump compresses the refrigerant vapour and this makes it get hot. As it flows through the condenser pipe it gives out heat and condenses back to liquid refrigerant. The liquid then passes through a small hole or expansion valve into a larger evaporator pipe. The lower pressure causes the liquid to evaporate and take in heat from its surroundings. Now the refrigerant vapour returns to the compressor and the cycle can start again. A thermostat controls the temperature by turning the compressor motor on and off as necessary.

▼ In a compression refrigerator, a compressor pumps the refrigerant round. The liquid refrigerant turns to vapour when it passes through the expansion valve. This produces a cooling effect. The vapour turns back to liquid in the condenser.

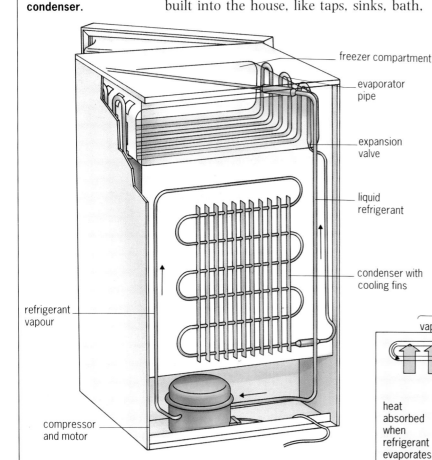

◀ The condenser pipe is outside the refrigerator, so it can lose heat to the air. The evaporator pipe is inside the refrigerator, so it can take heat away from the food.

Vacuum cleaners

A vacuum cleaner picks up dust and dirt from furniture and floors. Its motor turns a fan, making a partial vacuum inside the cleaner. The air outside rushes in, bringing with it the dust and dirt near the opening. This is all sucked up a tube into a bag, where the dirt is trapped while the air escapes. Some cleaners are pushed across the floor and have a spinning brush which loosens the dirt so it can be picked up easily. Others suck up the dirt with a nozzle at the end of a long, bendy tube.

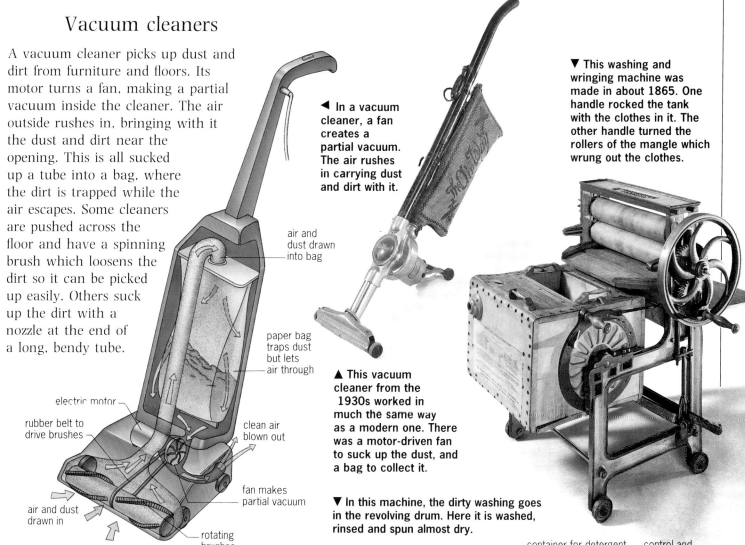

◀ In a vacuum cleaner, a fan creates a partial vacuum. The air rushes in carrying dust and dirt with it.

▲ This vacuum cleaner from the 1930s worked in much the same way as a modern one. There was a motor-driven fan to suck up the dust, and a bag to collect it.

▼ This washing and wringing machine was made in about 1865. One handle rocked the tank with the clothes in it. The other handle turned the rollers of the mangle which wrung out the clothes.

Washing machines

Just putting dirty clothes in water will not get them clean. Water must be forced through the fibres of the cloth. Most washing machines work by swirling the clothes in water in a revolving metal drum. Detergent is used to get the water more deeply into the fabric. It also removes the grease. Most washing machines are automatic. They can be left to do the whole wash by themselves. Often, they are controlled electronically. You choose from a range of different programs so that the machine hot-washes cotton clothes or cool-washes delicate woollens. It mixes hot and cold water to the right temperature, gives the clothes a long or short wash, rinses them, then spins out most of the water. Some machines tumble-dry as well.

▼ In this machine, the dirty washing goes in the revolving drum. Here it is washed, rinsed and spun almost dry.

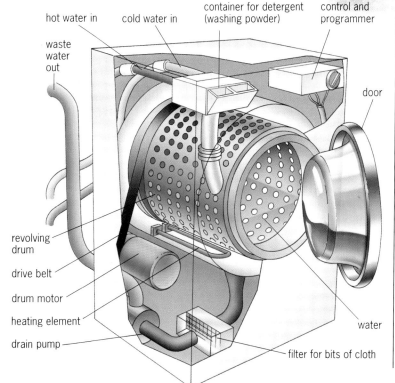

Cookers

Some cookers burn gas. Others use electricity or solid fuel. Gas cookers are particularly easy to use because they heat up instantly and cool down very quickly.

Hobs and ovens

The hob provides a number of areas of heat on which to boil, steam or fry food. On a gas hob the heat source is the flame of the burning gas. On an electric hob the heat is provided by a coiled element which is covered in metal or ceramic glass. Some electric hobs use heat from tungsten halogen lamps instead of elements. One of the newest types of electric hob is the induction hob. This works by creating a magnetic field. Food placed in a metal container in the magnetic field will heat up. With this method only the food gets hot.

The oven is an enclosed method of cooking. In most ovens the temperature at the top of the oven is higher than that at the bottom. Some electric ovens have fans which stir the air to keep the temperature the same throughout the oven. Some cookers have combined microwave and gas or electric ovens.

Microwave ovens and other small cookers

A variety of smaller electric ovens have been developed for specific purposes. Some, such as microwave ovens, are extremely compact and will fit into very small spaces. Microwave ovens can do most of the things the conventional cooker does whilst cooking food more quickly and so saving energy. Heat in a microwave oven is produced when the microwaves enter the food. These cause the food molecules to jump about. The resulting friction between the molecules brings about a rise in temperature, cooking the food from the inside.

Other small cookers, such as electric woks, can be used at table, and some, such as pressure cookers and high energy humidity cookers, speed up the cooking times. Slo-cookers are designed to cook very slowly, and deep fat friers and infra-red grills can be precisely controlled to do a particular job.

▼ This oven, set into the wall at eye level, has a glass panel and a temperature probe.

▶ A 19th century kitchen range set into a large alcove at Erddig House, Wrexham, Wales. Heat was produced by wood or coal burning in the central grate.

The first gas cookers were manufactured in 1834, but they were not widely available to buy until after 1860.

The first electric oven was manufactured in the USA in 1891.

Cooking

Almost all food can be eaten without any cooking at all. Most people eat raw fruit, and raw vegetables in salads, and the Japanese particularly like raw fish. However, some foods are much softer and easier to chew if they have been cooked. They may also be easier to digest. Thorough cooking also kills harmful bacteria that might cause food poisoning. Cooking often involves combining many different ingredients together. This can make food look and taste much better. Compare the taste of a chocolate cake, for example, with that of its raw ingredients of flour, sugar, fat, cocoa and egg. Different ideas about which foods to cook together have contributed to the very different dishes of the world.

What happens when we cook food?

Cooking causes chemical changes to take place in all food. These changes are irreversible. You cannot unboil an egg or turn that chocolate cake back into its original ingredients. The changes include softening of the cell walls and a tendency for the cells to separate. This can be seen if you look at mashed potato through a strong lens. The contents of the cells are no longer so well protected and some of them may escape. Water, for example, will evaporate during dry roasting, and a roasted joint of meat will shrink as it cooks. Colours and nutrients may be lost if food is boiled. So you should save vegetable water for soups and sauces.

Starch granules in food swell on cooking and become quite jelly-like. This is why flour is used for thickening. The protein in egg sets, and so eggs are used to hold and set dishes like soufflés and sponge cakes.

Heat works on food from the outside in. You can see this if you grill kebabs too close to the heat. They will burn on the outside before the centre is cooked. Microwave cooking heats with radio waves which penetrate deeper into the food.

The discovery of fire led to the invention of cooking. Prehistoric people roasted pieces of meat and fish on sticks over open fires and used flat stones placed by the fire to bake coarse flat bread. These stones were also used to heat the water in wooden or leather containers to make soup.

Some common cookery terms

Baste brush food with melted fat
Beat mix ingredients together vigorously with beater or spoon
Braise brown meat and then simmer in a little liquid in a closed dish
Cream mix ingredients with spoon or mixer till creamy
Dice cut into cubes
Fillet remove bones
Fold add ingredients gently with spoon to mix in without beating or stirring
Marinate soak food in spicy or herbal mixture to improve flavour
Poach simmer food in liquid to cook
Sauté fry gently until golden
Whip beat ingredients rapidly till foamy

boiling
The food is surrounded by water boiling at 100°C. If only a little liquid is used, and kept just below boiling point, this is called **poaching.**

steaming
The food is heated by steam from water boiling beneath it at 100°C. **Pressure cooking** is a type of steaming which uses pressurized steam at up to 120°C.

frying
The food is heated in hot fat at up to 190°C. It may be a coating of fat (as in stir frying), shallow fat, or deep fat.

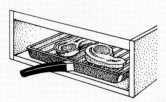

grilling
Radiant heat is used to heat the food, with temperatures reaching 200°C or more. **Toasting** and **barbecuing** work in the same way.

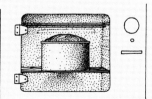

baking
The food is heated in an oven using hot air at up to 250°C. When meat is cooked in this way it is usually called **roasting.**

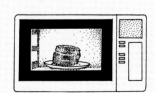

microwaving
Very short wavelength radio waves penetrate and heat the food.

◀ Different methods of cooking use different ways of getting the heat to the food.

Central heating

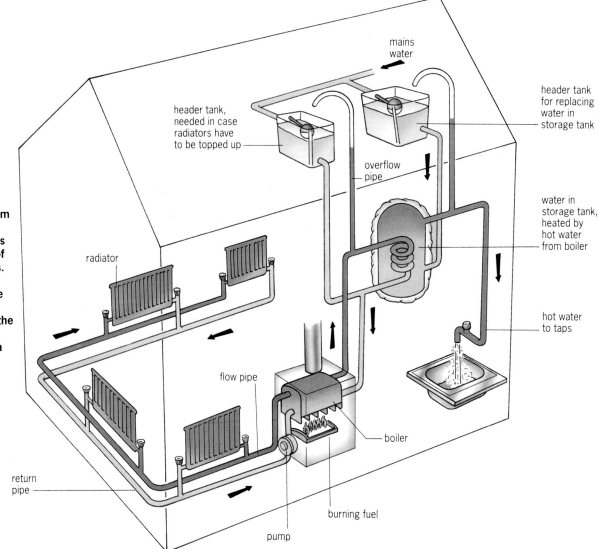

▶ A central heating system. Hot water from the boiler heats the radiators. It also heats the separate supply of water for the hot taps. The same water circulates through the boiler over and over again. This prevents the boiler and its pipes from scaling up like a kettle.

Central heating means that a whole building is heated by hot water or air from a single heat source. Most systems use water, because it is a very good material for storing and carrying heat. The water is heated in a boiler fuelled by gas, oil, electricity or solid fuel. It is circulated through radiators by an electric pump. The radiators heat the rooms. An adjustable valve at the base of each radiator controls the flow of hot water. The whole system is controlled by an electric timer. This automatically turns the heat on and off at the times you choose.

Hot water from the boiler is also used to heat the water for washing and baths. Some hot water from the boiler flows through a coiled tube in a storage tank. This heats the water in the tank, which is connected to the hot taps and shower. A pump is not needed for this purpose. Hot water rises naturally from the boiler. When it cools, it sinks back down to the boiler to be heated again. The same water circulates through the boiler over and over again. If fresh water were used each time, the system might scale up like a kettle.

Insulation

We insulate things to keep electricity, heat or sound where we want it. Wires have insulation to stop electricity flowing out if they touch against something. Ovens are insulated to keep heat in and freezers to keep heat out. Cars are insulated so that passengers do not hear the full noise of the engine.

Electrical insulation

Electrical wires can be insulated by covering them with a flexible plastic called PVC. In a mains cable, the insulation has two jobs to do. First, it stops electricity 'short circuiting' across to other wires. Second, it stops electricity flowing through somebody if they touch the cable.

Heat insulation

Our homes need insulation so that they can be kept warm with less fuel. Our bodies need insulation to survive when it is cold.

Trapped air is a good insulator. Clothes, especially woollen ones, trap tiny pockets of air in the fabric. They also keep a layer of air next to the body. Insulating materials used in the home rely on fibres, plastic foam or plastic beads to trap air. In double glazing, two sheets of glass have a layer of air sandwiched between them. A vacuum is an even better insulator than air. A 'vacuum' flask is a 'double-glazed' container with much of the air removed from between the inner and outer layers.

Sound insulation

Trapped air can also act as a sound insulator. Double glazed windows insulate against sound as well as heat loss. So do the cavity wall fillings used in modern buildings. Insulating can reduce the amount of traffic or aircraft noise getting into a building or the sound of loud music getting out.

Cars are carefully insulated to keep down the noise from their engines. Engines are mounted on rubber blocks so that less sound is transmitted to the rest of the car. Baffles in the silencer absorb sounds in the exhaust pipe.

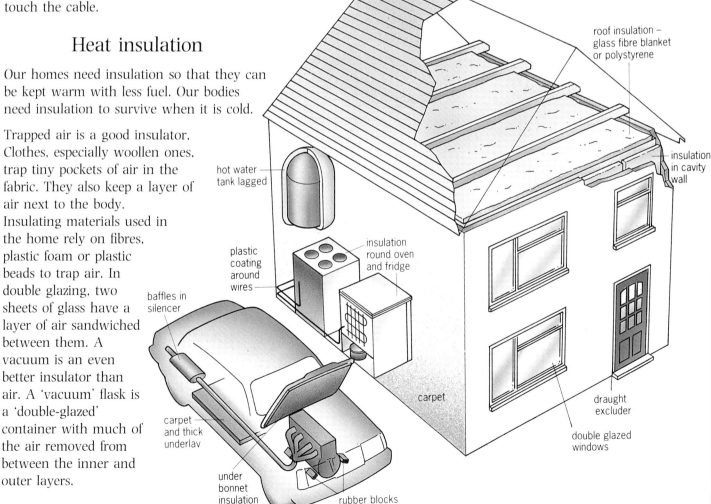

▼ Houses are insulated to reduce heat losses and, sometimes, to keep out unwanted sounds as well. Some items inside a house are also insulated to stop heat or electricity going where it is not wanted.

Transport

Humans have always been trying to find better ways of moving themselves and their goods around. Nowadays, most countries are criss-crossed with road and rail networks. Ships carry enormous cargoes from country to country. Aircraft can carry passengers faster than the speed of sound and space shuttles can, at enormous expense, transport a very few people into space.

Road

A car is a convenient way of getting around if only a few people want to travel. It can take you exactly where you want to go. Buses and coaches are a cheaper way of carrying lots of people but they can be boarded only at set stops and certain times. In small countries like Britain, trucks (lorries) are the main method of transporting goods. The goods being carried are called freight.

For fast, long-distance travel, cars, trucks and coaches need motorways to travel on. Some of these have three or more lanes in each direction.

Rail

Railways are the main freight carriers in large countries like the USA where huge distances have to be covered. For passengers, high-speed trains are a convenient way of travelling between large cities. They avoid tiring journeys on crowded roads and the problem of finding somewhere to park when you arrive. For those who want to take their vehicles, motorail services carry cars as well as passengers. Most cities have stations near the centre and many people use local trains to get to work. Building railway lines across cities has always been difficult. In some cities, the problem has been solved by putting the lines underground.

TRANSPORT TECHNOLOGY

Air

Flying is easily the fastest way to travel long distances. A cruise liner can sail across the Atlantic in three and a half days. Concorde, flying supersonic, can cross it in three and a half hours. One problem with flying is that airports are often a long way from city centres. Completing your journey by car or bus can make the total travel time much longer. Some cities have fast rail or helicopter links into the city centre. Others are building airports much closer in. These are designed for new, quiet aircraft which can take off and land from short runways.

Flying is an expensive way of moving freight. Very little freight goes by air compared with land and sea.

THE FUTURE

Already, more than 700 million passengers travel by air every year. And the number could double within ten years. For faster travel, spaceplanes are being developed that can carry passengers at more than five times the speed of sound. But planes like this burn huge amounts of fuel and may harm the atmosphere. In the future, most land, sea and air vehicles are likely to be quieter, use less fuel and cause less pollution.

Sea

Ships carry nearly all the freight that has to travel overseas. Some, like supertankers, are designed to carry one type of cargo. Others carry containers which are packed with goods before they reach the docks and can fit straight on the back of a truck.

Most people use ships only for ferry crossings. Many ferries carry vehicles as well as passengers. Ro-ro (roll on, roll off) ferries are designed for quick loading and unloading of vehicles. Hovercraft are used on some ferry services. They float on a cushion of air and can travel faster than ships.

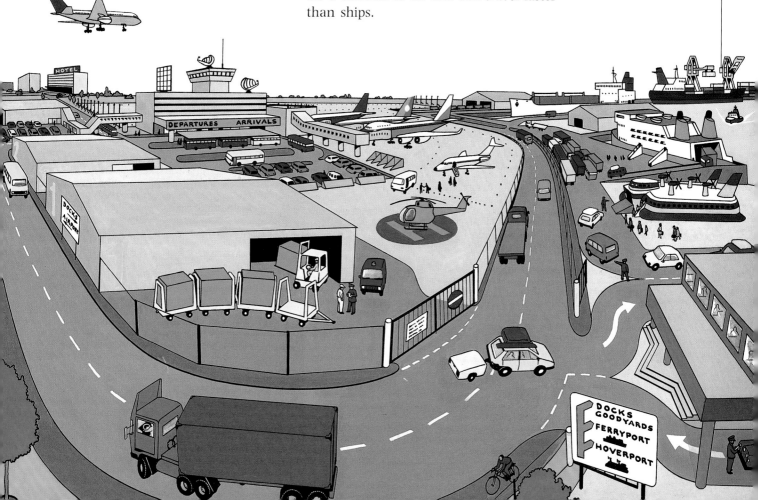

Flight

345 million years ago
First insects flew.
150 million years ago
First birds
50 million years ago
First bats
1783
Hot-air balloon carried man.
1853
Cayley's glider carried man.
1903
Wright brothers' *Flyer I* flew for 12 seconds.
1909
Blériot flew English Channel.
1919
Alcock and Brown's flight across Atlantic.
1941
Whittle's jet-engined aircraft.
1947
Bell X-1 flew faster than sound.
1977
First human-powered aircraft, *Gossamer Condor*.

People did not invent flying. Insects were flying nearly 350 million years before the first person dreamed of taking to the air. Insects stay in the air by beating their wings quickly. About 200 million years later, the first birds developed the ability to fly. They are well adapted to flying and have wings which are covered in feathers. To flap their wings, they have powerful muscles attached to a strong breastbone. Their bones are hollow, so their skeleton is light. The first mammals to fly were bats. They have wings made of skin stretched across their arm bones.

Getting into the air

If you watch a bird taking off you may see it jump into the wind and beat its wings. As the wing moves down, it pushes against the air and this produces an upward force which is bigger than the bird's weight. The air is also pushed backwards by the wing so that the bird moves forward. As the wing comes up, it is twisted and air passes between the feathers to cut down resistance.

Staying in the air

Air moving over a bird's wing creates lift. You can see this if you blow very hard over a small sheet of paper held level at one end. The paper rises slightly. This is because the moving air above the paper causes the pressure there to be lower than that below and lift is created. A bird's wing is arched and thicker at the front than the back. So the air travelling across the top has to travel faster than the air below. This creates lift in the same way as you discovered with the paper. It is this lift on the wings that helps keep the bird in the sky.

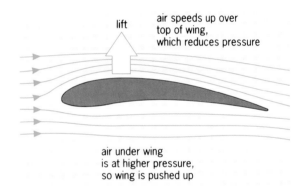

▲ Air flow across a wing.

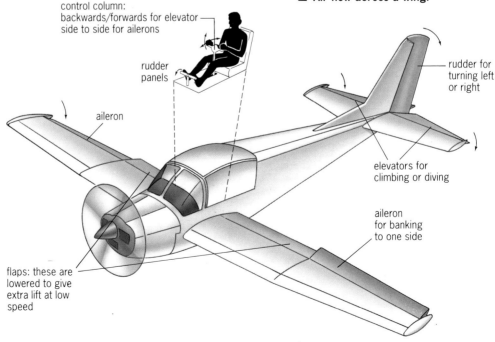

▶ The main control surfaces and controls used by the pilot.

TRANSPORT TECHNOLOGY

A bird tucks its legs up in flight so its body shape is smooth and streamlined. It uses its tail feathers to control which way it is going to fly.

The dream of human flight

Early attempts by humans to fly with artificial wings failed because no one had the strength to flap them quickly enough. In order to make a machine that would fly, inventors had to find a way of getting it into the air, keeping it there, and controlling its movement.

Getting an aircraft into the sky

Watching birds and experimenting with gliders helped early inventors understand that air moving over a wing can create lift. Most aircraft have wings with an arched cross-section like those of birds. The shape is called an aerofoil. But, unlike bird's wings, aircraft wings cannot be flapped. Therefore air must be kept flowing over them.

The aircraft in the picture is moved forward using a propeller that turns very quickly. The engine is similar to the one in a car, although more powerful. Modern airliners are moved using jet engines.

As the aircraft builds up speed, the air rushing over the wings creates the lift needed to raise it into the sky.

▲ A pilot controls a hang glider by pushing and pulling on a triangular-shaped frame. This pilot has a passenger as well.

Keeping it flying

The aircraft will stay in the sky as long as it keeps moving forward fast enough for the lift created to be the same as or greater than its weight. Aircraft designers have to ensure that the shape of the plane is streamlined like a bird in flight. This is because anything moving quickly through air is slowed down by air resistance. This force is called drag.

Controlling its movement

The pilot controls the movement of the plane using movable surfaces on the back of the wings, tail and fin. These are moved using a control column in front of the pilot's seat. To enable the aircraft to roll to the right or left, there are ailerons hinged on the back edge of the wings. These are connected to move in opposite directions. You can investigate their effect by making a simple paper aeroplane and cutting ailerons on each wing.

The aircraft can be made to climb or dive using the two elevators on the tailplane. These move in the same direction. If they are lowered, air pushes against the under-surface and the plane will dive. The pilot operates the rudder with two pedals. These control left or right movement.

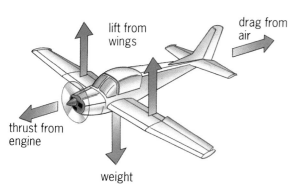

▲ The forces on an aircraft in flight.

Fastest bird
Peregrine falcon can reach speeds of 180 km/h in a stoop.

Fastest flying insect
An Australian dragonfly can reach 58 km/h in short bursts.

Fastest jet aircraft
The USAF Lockheed SR-71, a reconnaissance plane, reached 3,529 km/h over Beale Air Force Base, California in 1976.

Aircraft

Aircraft is the name we use for flying machines. Most have wings to keep them in the air but helicopters and hot-air balloons are really aircraft as well.

During the 19th century, short gliding flights were made in craft built by Sir George Cayley in England and Otto Lilienthal in Germany. However, it was not until 1903 that powered flight was achieved by the Wright brothers in the USA. In 1909, Louis Blériot flew across the English Channel. Ten years later, Alcock and Brown made the first non-stop flight across the Atlantic.

Passenger flying developed after World War I. During the 1920s, passengers often flew aboard mail planes, sometimes with the mailbags on their laps! By the late 1930s, flying was a much more luxurious affair. People could travel between Europe and the Far East aboard large flying boats which took off and landed on water.

The first jet aircraft, the Heinkel HE178, flew in 1939. However, jets were developed mainly after World War II. The first jet airliner, the De Havilland Comet, entered service in 1952. Today, nearly all long-distance international travel is by jet.

▼ The *Wright Flyer* makes the first-ever powered flight in 1903. Orville Wright is at the controls. His brother Wilbur is running alongside. The flight lasted just 12 seconds.

▲ The Handley Page HP42 entered service with Britain's Imperial Airways in 1931. It could carry up to 38 passengers. Safe and reliable, it had a maximum speed which was less than the take-off speed of a modern jet.

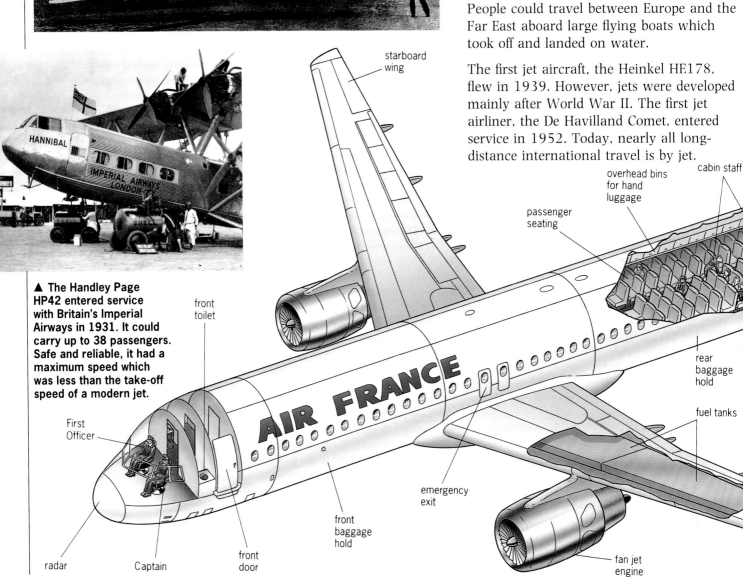

Shapes

Most aircraft have a central body called a fuselage, with wings near the middle and a smaller tailplane and fin at the back. Straight wings work best for carrying heavy loads at low speed, but swept-back wings give a better airflow for fast flying. Some military jets, such as the Panavia Tornado, have 'swing wings' which swing further back for high-speed flight. Some aircraft do not have a tailplane. Instead, the wings form a triangular shape, called a delta, which goes all the way to the back. Concorde is like this. Delta wings are good for high-speed flight but do not perform well at low speed. When Concorde lands, its wings are set at such a steep angle that the nose has to be lowered to give the pilot a forward view. A few aircraft have their wings at the back and their 'tailplane' at the front. This is called a canard arrangement.

Power

Most modern aircraft use jet engines in one form or another. Even where a propeller is fitted, the power may come from a turbo-prop engine which is based on the jet. Until the 1950s, most aircraft had propellers turned by piston engines. Some small aircraft still do. These engines work in the same basic way as a motor car engine.

The changing face of design

Concorde was designed in the 1960s. Nowadays, airlines are more interested in economy than speed, and other airliners travel at less than half Concorde's speed. Using computers, designers have developed wings which slip more easily through the air, and engines which are quieter and burn their fuel more efficiently. A 'jumbo jet' can carry four times as many passengers as Concorde using only the same amount of fuel.

Computers are an important part of a modern airliner. The autopilot is a computer which can navigate and fly the aircraft for most of its journey. On some aircraft, the pilot does not directly control the plane. Instead, the pilot's controls send instructions to a computer and the computer works out the best way to fly the plane.

◀ This Airbus can seat up to 179 passengers. It is designed for short and medium-range flights. The aircraft needs a flight crew of only two. It has computers to help manage its engines and controls.

Monoplane
An aircraft with a single set of wings. Most modern aircraft are monoplanes.

Biplane
An aircraft with two sets of wings, one above the other. Most aircraft built before the mid-1920s were like this. Some modern aerobatic aircraft are biplanes.

VTOL
Vertical Take-Off and Landing aircraft do not need runways because they can hover.

STOL
Short Take-Off and Landing aircraft are designed for slow-speed flying and short runways.

◀ A Concorde prototype makes its first flight at Bristol, England, in 1969. Concorde was the first supersonic (faster-than-sound) airliner to enter service. It can carry up to 144 passengers at twice the speed of sound and can cross the Atlantic in 3 hours.

Balloons and airships

► This hot-air balloon is 16 m across. It can carry three people in the basket underneath. To make the balloon rise, the pilot heats the air by using the gas burner in short bursts.

The balloon is made of nylon and can be as big as a house when it is inflated. A panel at the top is removed after landing to allow the hot air to escape quickly. The basket hangs under the balloon from wires or ropes. It carries the crew, and also the gas cylinder and the instruments, which the crew need to calculate their height, direction, and how much fuel they still have. The gas burner is used in short bursts to keep the balloon at the chosen height. Balloons go where the wind takes them. Experienced pilots change altitude to make use of the winds.

Over 200 years ago the first person to fly did so with a hot-air balloon designed by the Montgolfier brothers in France.

How do balloons fly?

Hot air is lighter than cold air and so it rises. To make a balloon fly, the air inside it must be heated. Modern balloons carry gas burners to do this. If the burner is turned on, the balloon rises. When the air inside cools, the balloon loses height.

The toy balloons at fairs are not filled with hot air but still rise up. They are filled with helium gas, which is lighter than air. Hydrogen is the lightest gas and was used in passenger-carrying balloons and airships until the 1930s. Unfortunately it catches fire easily and after a series of bad accidents it ceased to be used. In recent years balloons are again beginning to be used for transport but they are filled with helium, which cannot burn and is more readily available than it was in the 1930s.

Hot-air balloon records
Richard Branson (UK) and Per Lindstrand (Sweden) made the first transatlantic flight in 1987.

The duration record is held by Hélène Dorigny and Michel Arnould (France), who flew for over 40 hours non-stop in 1984.

In 1980, Julian Nott (UK) gained an altitude record when he reached 16·8 km (10·4 miles).

Don Cameron and Christopher Davey (UK) set a distance record in their gas and hot-air balloon, *Zanussi*. They travelled over 3,339 km (2,074 miles) in 1978.

Some uses of balloons

Balloons are used to collect information about the weather. These do not carry passengers but are loaded with instruments to measure atmospheric pressure, temperatures and wind speed. The readings are either stored on board until the balloon returns to the ground or sent back to earth by radio signals. They help meteorologists predict what the weather is going to be like. Astronomers and other scientists also use balloons to carry out experiments and collect data from high in the atmosphere.

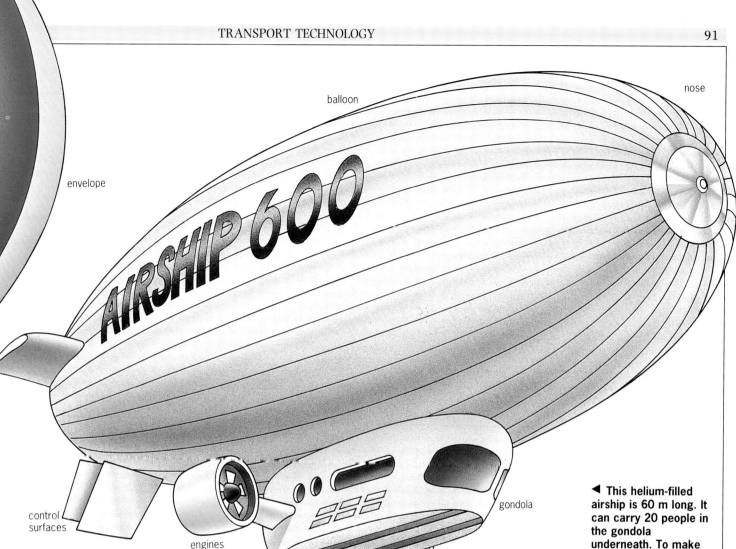

Airships

Airships, also called dirigibles, are sausage-shaped balloons powered by engines. Modern ones are filled with helium gas. They have an enclosed gondola below, which can carry as many as twenty passengers. The motors which turn the propellers to move the airship forward are attached to the gondola. Unlike other balloons, an airship is not dependent on the wind. The pilot can steer it using a large rudder. Like an aeroplane, it has elevators to make it climb or dive.

A journey in a modern airship is quiet, safe and comfortable. Because it can remain stationary in the air, it is ideal for photographic and television work. You may well have watched pictures of a sporting event filmed from an airship. Airships also make good flying cranes and can be used to lower heavy structures accurately into position. They are not much used for commercial passenger flights.

Flashback

In 1783, the Montgolfier brothers built the first hot-air balloon which carried a person. The air was heated using a fire on the ground. The first hydrogen-filled balloon flew in the same year. These balloons were often used for military observation. In 1849 balloons were used to bomb the Italian city of Venice. The first flight of an airship, *La France*, was in 1884. Many huge hydrogen-filled airships were built to carry large numbers of passengers. On 6 May 1937 the *Hindenburg* burst into flames when about to moor near New York, killing 35 of the 97 people on board. After that accident hydrogen-filled airships were not used again. Two Americans set an altitude record in 1961. Their helium-filled balloon reached a height of 34·7 km (21·5 miles). They had to carry their own air supply to breathe at that height.

◀ This helium-filled airship is 60 m long. It can carry 20 people in the gondola underneath. To make the airship climb or dive, the pilot alters the angles of the elevators ('wings') at the back of the engines.

In 1929 the *Graf Zeppelin* took ten days to travel completely around the world.

World's longest airship:
Zeppelin LZ129 *Hindenburg*
Built: Germany 1936
Length: 245 m
Crew: 61
Passengers: 36
Accommodation:
2 passenger decks (with single and twin berth cabins), dining room, reading room, writing room and lounge (with piano).
Cruising speed:
125 km/h (78 mph)
Time to cross Atlantic:
65 hours

Hovercraft

The French word for hovercraft is *aeroglisseur*.

The first hovercraft, the SRN1, flew from the Isle of Wight to mainland England in 1959. It was invented by the British engineer Sir Christopher Cockerell. The first passenger service using hovercraft began in 1965. This also went to and from the Isle of Wight.

Hovercraft carry people and goods over land and sea. They are not really boats or aircraft, but float on a cushion of air. They are sometimes called 'air cushion vehicles'. Hovercraft can travel over any fairly flat surface and move easily from water to land without stopping.

How it works

A hovercraft does not touch the surface over which it is travelling. It floats on a layer of air which is made by pumping air downwards underneath the hovercraft. This is kept in by large rubber skirts, but the engines have to work hard to keep enough air underneath to lift a large hovercraft. The hovercraft is pushed forward by propellers on top like an aircraft. Many hovercraft actually use ordinary aircraft engines. They are steered by turning huge fins or the propellers themselves.

Hovercraft in use

Some hovercraft are used as passenger and car ferries. The trip is called a flight, just as in an aircraft. Some of these hovercraft weigh over 160 tonnes and can carry 400 passengers and 55 cars. They are much faster than normal ferries because they do not have to push through the water. They cruise at about 80 km/h (50 mph) and fly with their skirts just above the water. Hovercraft are used in many places. In Canada they work across land, water and ice. Smaller hovercraft are used for fun and people have even made their own hovercraft. Other machines, like the hovermower that cuts the lawn, float on a cushion of air just like a hovercraft.

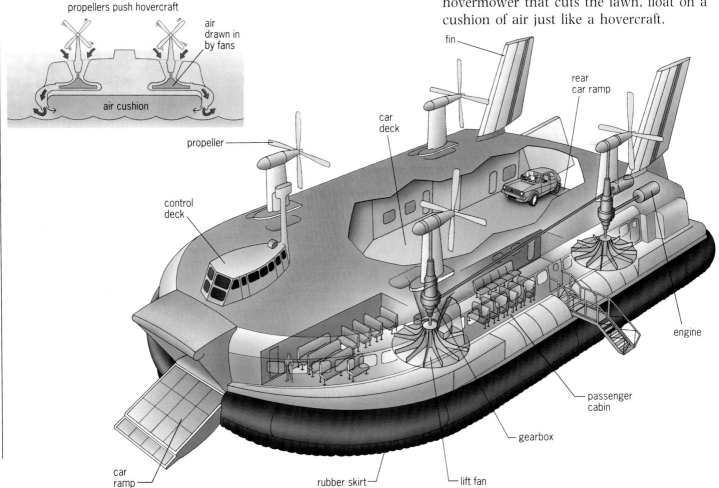

Helicopters

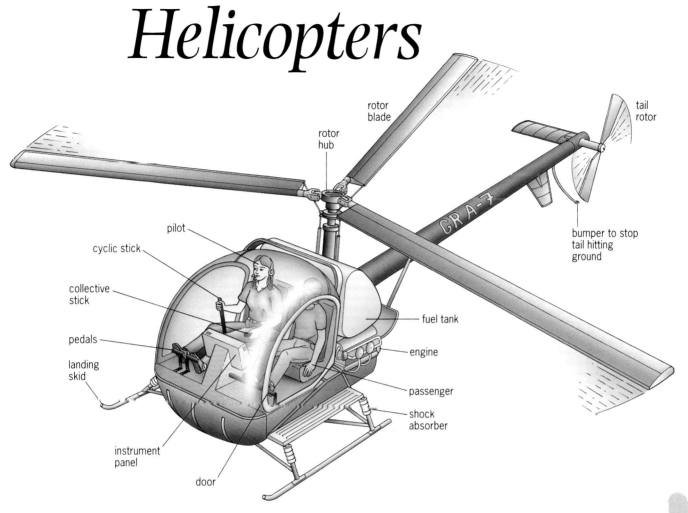

Helicopters fly, but not in the same way as ordinary aircraft. Instead of wings they have long thin blades on top which turn to lift the helicopter. They are smaller than aircraft and are noisy, but they can take off and land from any flat area, even a flat roof. So they are often used as air taxis to carry passengers short distances.

How they work

A helicopter has an engine which makes its blades turn. As they turn, they push air downwards and this lifts the helicopter upwards. By tilting the blades the pilot can make the helicopter take off, hover or land. To go forwards, the blades must be tilted so that they push some of the air backwards as well as downwards. Helicopters also usually have another small rotor on their tail. This is to stop it spinning round in the opposite direction from the main rotor. Helicopters cannot fly as fast as aircraft.

Types of helicopter

Most helicopters have one large rotor but some that are used to lift heavy weights have two or more. Another type of helicopter is the autogiro. It cannot take off straight up, because its rotor blades are turned not by its engine but by air rushing past them like a windmill. A propeller pushes it forward and the air turns the blades, lifting it.

Uses of helicopters

Helicopters are used to carry people and equipment to and from inaccessible places. They fly crews and equipment to offshore oil rigs and are used for short-distance journeys between airports and across towns. They carry soldiers quickly around the battlefield and work as gunships. They can fly from ships and are ideal for rescue work on mountains or at sea because they can hover.

▲ **Controls used in a helicopter.**
1 cyclic stick for forwards, backwards or sideways
2 pedals for turning left or right
3 collective stick for up and down, with twist grip for engine speed

First practical helicopter
The Focke Fa-61, designed by H. Focke (Germany) in 1937

Maximum speed
About 400 km/h (250 mph)

Jet engines

For fuel, jet engines use kerosene, a type of paraffin.

A Boeing 747 Jumbo Jet with four jet engines uses 180,000 litres of fuel to fly from London to Hong Kong. That amount of fuel would fill a car's tank 3,500 times and take the car over a million miles.

Typical jet engines
Rolls Royce RB211 Turbofan
Diameter 2·2 m
Length 3·0 m
Thrust, strong enough to lift 26 tonnes
Weight 4·5 tonnes
Used in Boeing 747 Jumbo jets

Pratt and Whitney F100-PW-100 Turbofan
Diameter 1·2 m
Length 5·3 m
Thrust, strong enough to lift 11 tonnes
Weight 1·5 tonnes
Used in modern jet fighters

The first jet engine was tested in Britain in 1937 by Frank Whittle. But the first jet aircraft to fly was the Heinkel He178 in 1939 using engines developed by von Ohain in Germany.

The first jet airliner was the de Havilland Comet which went into service in 1952.

Jet engines are very powerful engines used mainly to drive large, fast aircraft. They have fewer moving parts than the internal combustion engines used on earlier aircraft and need less looking after. The widespread use of jet engines on airliners in the 1960s made world-wide air travel faster and cheaper.

How they work

The jet engine works like a rocket. But unlike a rocket, it sucks air in at the front. In a basic engine, air is compressed and forced into a combustion chamber where fuel is added and burnt. This produces very hot gases which expand and escape out of the back of the engine. As the gases rush out, they drive the aircraft forward. They also push round the turbine which turns the compressor. For extra power more fuel can be squirted straight into the exhaust gases and burnt. This is called afterburning. It makes the gases even hotter, but uses lots more fuel. Afterburning is used in fighter planes when they need extra speed.

Types of engine

Turbojets were the first type of jet engine. All the air sucked into the engine passes through the combustion chamber. Turbojets are very noisy because the gases rush out of the engine at high speeds.
Turbofan engines are quieter because they push out air at lower speeds. They have huge fans at the front to collect air. Some goes through the combustion chamber, but most is pushed around the outside. Most large airliners use turbofan engines. In **turboprop** engines a turbine drives a propeller which pulls the aircraft forward. They work best at speeds below 800 km/h (500 mph) and are often used in helicopters. The **ram-jet** is the simplest jet engine. It does not suck air in but scoops it up as it flies along. Ram-jets must be moving fast before they can collect enough air to work.

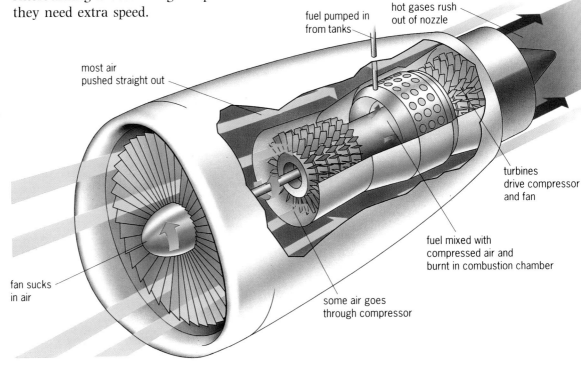

▶ A turbofan engine.

Steam-engines

Steam-engines were the first fuel-burning engines to be invented. During the 18th and 19th centuries, they were the main source of power for industry. With coal as their fuel, they drove machinery in factories and later powered ships and trains. Today, there are very few steam-engines left. Other types of engine have taken over.

How they work

Most steam-engines use the pressure of steam to push a piston up and down a cylinder. The steam is made by boiling water over a fire of burning coal or oil. Steam from the boiler is let into the top of the cylinder by a valve. The steam expands and pushes the piston down. When the piston reaches the bottom, the valve changes position. Now, the valve lets the first lot of steam escape and feeds fresh steam into the *bottom* of the cylinder. This pushes the piston back up. When the piston reaches the top, the valve changes position again, the piston is pushed down . . . and so on. The up-and-down movement of the piston is turned into round-and-round movement by a crank. The crank moves a heavy flywheel which keeps the engine turning smoothly. The flywheel can drive other wheels or machinery.

Steam-engines work best with high-pressure steam. To withstand the pressure, the boiler, pipes and cylinders have to be very strong. In some engines, there is a condenser to collect and cool the escaping steam. The steam condenses (turns into water), goes back into the boiler and is used again. With a condenser, a steam-engine can work for longer without running out of water.

Flashback

The first practical steam-engine was built by Thomas Newcomen in England in 1712. It used to pump water from mines. In 1785 James Watt improved on Newcomen's design, and in 1962 he invented his sun-and-planet gear, a type of crank which turns up-and-down movement of a piston into rotation. Steam-engines could now drive machinery. Watt's engines provided much of the power for Britain's mills and factories during the Industrial Revolution. In 1804 Richard Trevithick built the first steam locomotives, and later steam-powered barges.

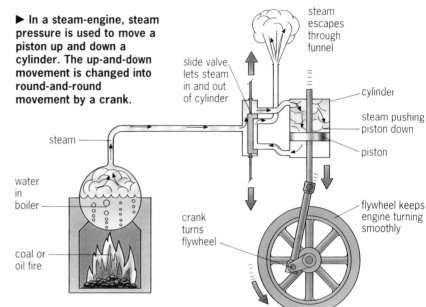

▶ In a steam-engine, steam pressure is used to move a piston up and down a cylinder. The up-and-down movement is changed into round-and-round movement by a crank.

▼ This threshing machine is driven by a steam-engine. Engines like this were used on farms in Britain and North America in the second half of the 19th century.

Trains

▲ Union Pacific's 'Challenger'.

Today the word locomotive means a railway engine, but many years ago it also meant a road engine. The earliest locomotives were driven by steam, but nowadays most railways use diesel or electric locomotives.

Diesel and electric locomotives

Diesel locomotives have engines similar to those in trucks and buses. However, in most large diesel locomotives the engine does not drive the wheels directly. Instead, it turns a generator which supplies electricity to electric motors mounted between the wheels. Locomotives that use this system are sometimes known as diesel-electric locomotives. The system is a good

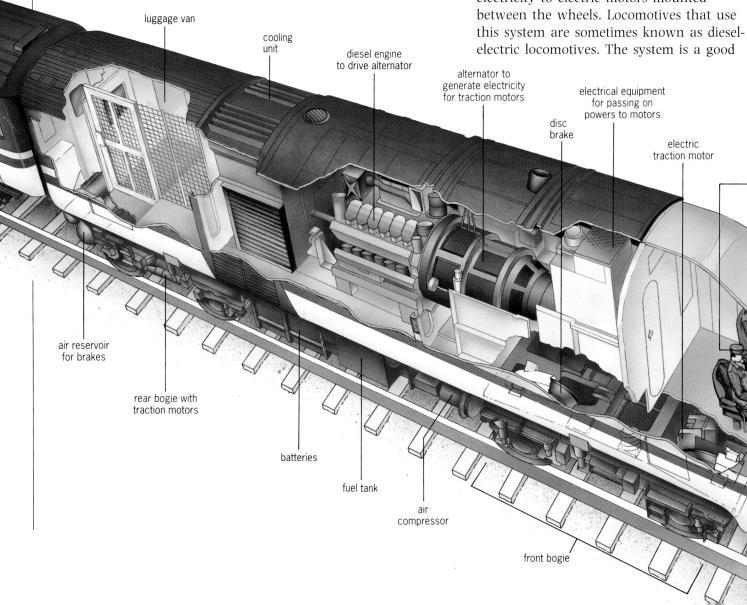

way of applying smooth power to the wheels without the need for any gear-changing, but it works efficiently only with very large powerful engines. Local services often use trains known as diesel multiple units (d.m.u.s). In these, small diesel engines are mounted under the floor of each coach, and there is gear-changing just as on a truck or bus. With a d.m.u., there is no need for a separate locomotive.

Electric locomotives can be small but powerful because they do not have to carry a large engine and a supply of fuel. Electric motors turn the wheels. The electricity is supplied from wires suspended above the track or from an extra, third rail alongside the track. Electric multiple units (e.m.u.s), like d.m.u.s, have no separate locomotive.

Trains of the future

High speed trains of the future will have no wheels. Aerotrains are held away from the track by an air cushion, rather like a hovercraft, and propelled by jet engines. Maglev (magnetic-levitation) trains are kept in the air and propelled by magnetic forces.

—— assistant
—— driver
—— controls

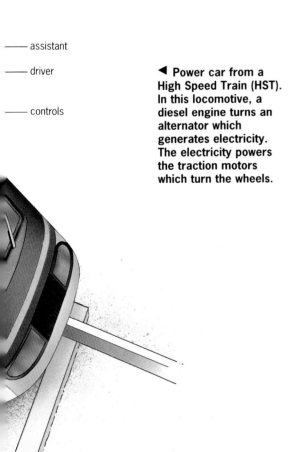

◀ Power car from a High Speed Train (HST). In this locomotive, a diesel engine turns an alternator which generates electricity. The electricity powers the traction motors which turn the wheels.

GEORGE STEPHENSON

George Stephenson's father was a colliery fireman, and as a teenager George had various jobs working with mining engines. He had never been to school, but he taught himself to read by attending night-classes. He became a colliery engineer, and in 1812 was appointed as engine-builder at Killingworth colliery, near Newcastle. In 1815, Stephenson invented a miner's safety lamp. But Sir Humphry Davy, who produced a similar lamp at the same time, is generally regarded as the safety lamp's inventor.

Stephenson spent most of his working life designing and building railways and railway locomotives. Between 1814 and 1826 he built at least twelve railway engines for pulling coal, but in 1823 he was put in charge of building a railway from Stockton to Darlington that would carry people. When it opened in 1825, Stephenson himself drove the engine which pulled the world's first steam-hauled passenger train. From 1826 to 1830 he supervised the building of the Liverpool to Manchester railway. Before it opened, a competition was held to find the most efficient locomotive. The £500 prize was won by 'The Rocket', designed by Stephenson and built by his son Robert.

Stephenson's successes led to the building of railways throughout Britain. From 1830 until he retired in 1845, he acted as engineering consultant to several railway companies. He used the money he had made from his inventions to set up schools for miners' children and night-schools for the miners themselves.

George Stephenson's fame has tended to overshadow the achievements of his son Robert, who assisted his father as railway-builder and locomotive-designer and became a famous bridge-builder.

▲ 'The Rocket' was built by George Stephenson's son Robert. The front wheels were turned by steam-driven pistons. Coal and water for the boiler were carried in the truck on the back.

Railways

▲ Blackfriars Bridge Station, London, where thousands of commuters arrive daily to work in the City of London.

The smooth metal tracks of railways allow heavy loads to be moved more efficiently than on roads. A single locomotive may pull a load weighing thousands of tonnes. The track it travels on is made of steel rails laid on concrete or wooden sleepers. The sleepers lie on a layer of rock chippings. These help to take the weight of the passing train and keep the rails level. The two rails are laid a certain distance apart, known as the gauge. The most common railway gauge is 1.44 m (4 ft 8½ in). Finland and Spain use a wider gauge.

Constructing railways

When a railway is to be built, engineers choose a route that is as level as possible. Any slope or gradient has to be very gradual, or the steel wheels of the locomotive will slip. Curves also have to be as gentle as possible, because a train cannot turn sharply at high speed. Viaducts are used to carry the railway across valleys, and tunnels to carry it through hills.

In mountainous countries where the slope of the track is very steep, a special extra rail is used between the outer rails. This is called a rack rail. It has teeth that point upwards, and a powered gearwheel on the locomotive, called a pinion, engages the teeth. This allows the train to pull itself up a hillside without slipping.

When a track is laid, short gaps are often left between the lengths of rail. This allows the rails to expand when they warm up in summer. Without the gaps, the rails would bend out of shape. Sometimes extra long rails are used. These are fixed very tightly to the sleepers so they cannot expand in the heat. Railway tracks are regularly checked, often by trains with measuring instruments on board.

▶ Train wheels have flanges to keep them on the rails. The rails are mounted on sleepers which are embedded in stone chippings.

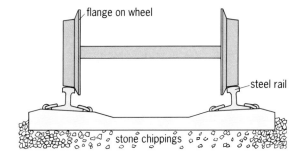

Signalling

To avoid collisions, a railway line is divided into sections with a signal at the beginning of each section. Signals are like traffic lights. A train can pass a signal only if it shows the right colour or the right position.

The signals are operated from a signal box. In modern signal boxes, the positions of all the trains on the line are displayed by a computer. The computer ensures that only one train is travelling on each section of line.

The signal box also controls the route that each train follows. There is no steering-wheel in a locomotive. Instead, trains are steered by special moving rails, called points. These are used to make a train change direction where tracks divide or where they join together.

Types of train

The fastest modern trains can travel at speeds of over 300 km/h (200 mph). They are powered either by overhead electric wires or by diesel engines. High-speed passenger trains are widely used in Europe and Japan. Engineers are experimenting with trains that use magnets to make them 'float' above the track by magnetic levitation.

The load that a locomotive pulls can be either passengers or freight. Commuter trains carry people to and from their work, transporting far more people than could be carried in cars or buses. Freight trains carry cargo. They are often manufactured specially for carrying such substances as oil, chemicals, clay, cars and iron ore. Some freight trains can be over 500 m (1,600 ft) long.

Flashback

The first passenger railway opened between Stockton and Darlington, England, in 1825. The Liverpool to Manchester railway opened five years later and public railways opened in Belgium and Germany in 1835. During the mid-19th century railways expanded very rapidly in England and throughout the industrialized world.

In the early days of railway building, all the cuttings, embankments and tunnels were dug by hand. Teams of thousands of men, using pickaxes and other hand tools, worked to move the earth and lay tracks.

Railways brought great changes to the way people lived. For the first time in history people could travel long distances every day to work. This meant that cities could grow bigger, with outer suburbs connected by train.

In 1825 a locomotive reached a speed of 24 km/h (15 mph).

By 1938 the speed record for a steam locomotive was 203 km/h (126 mph).

The French Train Grande Vitesse (TGV) reached a speed of 518 km/h (322 mph) in May 1990.

The Canadian Pacific line was completed in 1871. The Trans-Siberian Railway, from Moscow to Nakhodka, is the longest in the world at 9,438 km (5,865 miles). There are 97 stops on a journey which takes just over 8 days.

The first underground railway in the world was the London Underground, which opened in 1863. Today it has 409 km (254 miles) of tracks and 272 stations. Almost 800 million passengers use it every year.

▲ A heavy freight train on the Alaskan Railroad being pulled by five locomotives.

▼ A French TGV express train.

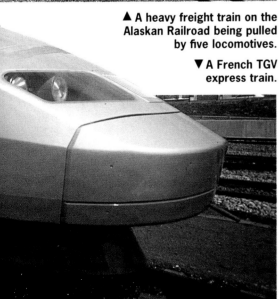

◄ This lithograph shows Edgehill Station, Liverpool in 1848. Luggage is being strapped onto the roof of a train.

Cars

The world's largest car is a Cadillac, specially built for its owner in the USA. It is over 18 m long. It has sixteen wheels, and a swimming pool in the back.

There are more than 400 million cars in the world today. Most have four wheels, though some have only three, and at least one has sixteen!

Saloon cars have a closed-in compartment for passengers, with a separate boot (trunk) for luggage. Coupés are saloons designed for two people. Estate cars (station wagons) are like vans with extra windows. They have a door at the back and rear seats which fold down to give more luggage space. Hatchbacks are like saloons, but with folding seats and a rear door like estates. Sports cars are designed for fast driving. Most carry only two people. Some have fabric roofs which fold down. GT (Grand Touring) cars are saloons which perform like sports cars.

Safety

Modern cars are designed with safety in mind. The passenger compartment is built to resist collapse in an accident, while the front and rear sections are designed to crush up. This makes a collision less violent for the people inside. There are seat-belts to stop passengers being thrown through the windscreen and special locks which stop the doors bursting open. The windows are made of safety glass which has no sharp edges when it breaks, and the steering column is collapsible in case the driver is thrown against it. Some cars have bars in the doors to protect passengers in side-on collisions. And some have antilock brakes to make skids less likely.

Flashback

The first petrol-driven cars were made in Germany in the 1880s. Passengers sat out in the open and, in Britain, someone had to walk in front waving a red flag. Early motoring was expensive. But, in 1907, Henry Ford in America started to produce a cheap reliable car, the Model T. In the 1920s, a wide choice of cheap cars became available and the closed-in saloon body became popular. Since then, there have been many improvements but the basic design of the car has changed very little.

▶ Carl Benz driving a 1888 Benz three wheeler.

◀ A 1953 Volkswagen Beetle. The Beetle was first produced in 1938; by 1981, 20 million had been made.

▼ The Austin Mini, designed by Alec Issigonis, first went on sale in 1959 for £496 19s 2d.

▼ 1957 Cadillac Eldorado Biarritz.

How a car works

A car has over 20,000 parts, grouped into four main sections. The body holds the driver and passengers. Attached to this, there is an engine, a transmission system and a set of controls for the driver. The engine is the heart of the car. It runs on petrol or diesel fuel and provides power to drive the wheels. The transmission system takes the power from the engine to the wheels. For the driver, there is a steering-wheel to point the car in the right direction, an accelerator to make it go faster and brakes to slow it down. The driver may also have a clutch and a gear-lever for changing gear, though in some cars this is done automatically.

Instruments for the driver

Speedometer tells you the speed of the car in km/h or mph.

Odometer tells you the total distance travelled by the car in kilometres or miles.

Tachometer (rev counter) tells you the engine speed in revolutions per minute (rpm).

Suspension

This joins each wheel to the car body. It allows the wheels to go up and down over bumps while giving the passengers a smooth ride. Most suspensions use coil springs. They have dampers (shock absorbers) to stop the car bouncing up and down.

Steering

The car is steered by turning the front wheels in the direction you want to go. The steering-wheel is connected to the wheels by a steering-box and two rods which move from side to side.

Engine

This provides the power to drive the car. It is normally an internal combustion engine, burning petrol or diesel fuel. You push down the accelerator to make it go faster. Some cars have the engine at the rear.

Electrics

A battery provides electricity for the starter motor and spark-plugs when starting the engine. It also powers the lights and the radio when the engine is not on. When the engine has been started, it drives an alternator. This supplies the electricity for the lights, plugs and other equipment. It also recharges the battery.

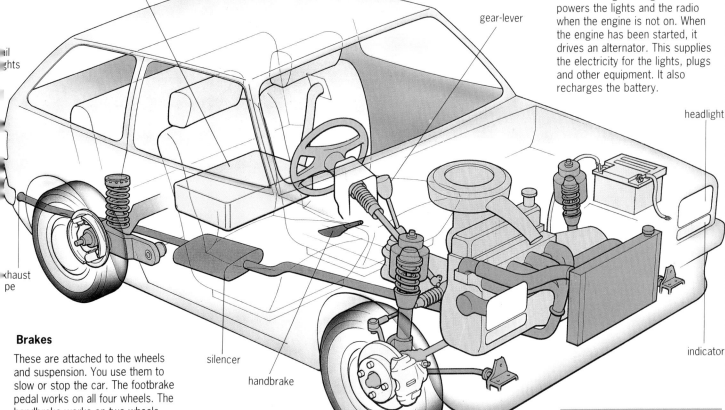

Brakes

These are attached to the wheels and suspension. You use them to slow or stop the car. The footbrake pedal works on all four wheels. The handbrake works on two wheels and is used for parking. There are two kinds of brakes: disc brakes and drum brakes. Most cars have disc brakes at the front and drum brakes at the back.

Gearbox and clutch

Most cars have four or five forward gears and one reverse gear. The low gears, called first and second, are used for starting off and driving up steep hills. The higher gears are used for normal driving. To use the clutch, you press a foot pedal. It disconnects the engine from the gearbox while you change gear.

Driveshafts

These are metal rods (one either side) which transmit the turning force from the engine and the gearbox to the wheels. Most small cars are front wheel driven like this one. The back wheels just roll along. Larger cars often have rear wheel drive. In some cars, all four wheels are driven.

Cooling system

When the engine is running it gets very hot. To stop it overheating, the cooling system pumps water around the engine and then through the radiator. The hot water in the radiator is cooled either by a rush of air passing through as the car drives along or by a fan when the car is not moving.

Motor bikes

▶ This trail bike is designed to be ridden over rough country tracks.

Labels: fuel tank, kick-start, carburettor, twist-grip throttle, speedometer, telescopic front forks with dampers and springs, silencer, knobbly tyres for good grip on rough ground, rear brake (drum brake), front brake (drum brake), gearbox, engine, exhaust pipe, sprung swing arm to allow rear wheel to move up and down

Motor cycles with an engine capacity of less than 50 cc are called mopeds. At one time, machines like this had pedals as well as an engine.

Motor cycles have a steel frame rather like that of a bicycle. The engine, gearbox, fuel tank, saddle and other parts are all bolted to this frame. Front and rear wheels are sprung and have hydraulic dampers (shock absorbers) to stop the bike bouncing up and down too much. The engine is connected to a gearbox which turns the rear wheel, usually by means of a chain. Engines range in capacity from less than 50 cc to over 1200 cc. Small motor cycles usually have air-cooled, two-stroke petrol engines. Larger and more expensive machines usually have four-stroke engines, which can give better fuel economy. They may also have water cooling. Many motor cycles have electric starters to get the engine going, though kick-starts are common on smaller machines. Front brake, throttle, clutch, and lighting controls are mounted on the handle-bars. The rear brake is applied by a foot pedal. Gear changes may also be made by foot. However, many motor cycles have automatic gear boxes.

Flashback

The first motor cycle was a bicycle powered by a small steam-engine. It was constructed by Ernest and Pierre Michaux in 1868. Gottlieb Daimler built the first petrol-engine motor cycle in 1885. At one time, motor cycles were made all over the world, but today, most are built by Japanese companies.

▲ 1923 Bohmerland 600cc motor cycle with side-car, made in Germany.

▼ 1903 Triumph, made in Britain.

Internal combustion engine

Cars, motor cycles, trucks and buses are all powered by internal combustion engines. Most use petrol or diesel oil as their fuel. The fuel is burnt in a series of explosions. The force of each explosion pushes a piston down a cylinder. The movement is used to turn the wheels of the vehicle.

How a petrol engine works

The diagrams below show how a single-cylinder petrol engine works. The engine is called a four-stroke engine because it keeps repeating a series of four up-and-down movements or strokes. During the power stroke, a mixture of petrol and air is exploded by a spark from a plug; then burnt exhaust gases are removed and a fresh mixture of petrol and air is drawn in and compressed, ready for explosion. The up-and-down movement of the piston is turned into round-and-round movement by a crank fixed to a heavy flywheel.

Many petrol engines have four cylinders, with their pistons all turning the same crankshaft. Each cylinder is on a different stroke. This helps the engine run smoothly. The cylinders get very hot when the engine is working. Usually, they are cooled by pumping water round them.

Most cars and many motor cycles have four-stroke petrol engines with two, four, six or more cylinders. Some motor cycles and lawn-mowers use two-stroke engines. These have only a compression stroke and a power stroke. The exhaust gases are pushed out by the force of the fuel rushing in.

Diesel engines

Most trucks and buses use four-stroke diesel engines, as do some cars. A diesel engine does not need spark plugs. The compressed air gets so hot that the oil starts to burn as soon as it meets it. Diesel engines are less powerful than petrol engines but they use less fuel, and have less harmful exhaust gases.

Carburettor
Mixes petrol and air in the right proportions for an explosion.

Ignition coil
Turns low-voltage electricity from the battery into the high-voltage electricity for the spark plugs.

Distributor
Sends sparks to the plugs in the right order.

Camshaft
Is a spinning shaft with cams (bumps) on it to push the valves open.

Radiator
Gets rid of the heat from the water which cools the engine.

Fan belt
Drives the water pump and the alternator (for charging the battery). In some cars, it also drives the fan which cools the radiator.

Other fuels
Some petrol engines have been converted to run on alcohol, camping gas and gas made from manure.

Engine capacity: one litre
This means that, between them, the cylinders can suck in one litre of air and fuel mixture when their pistons go down.

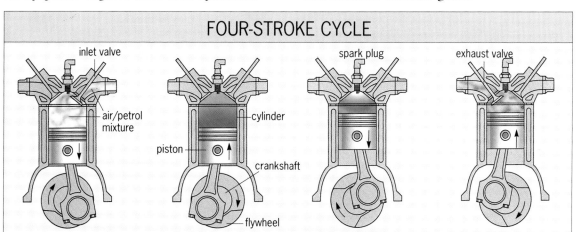

FOUR-STROKE CYCLE

Induction stroke
The inlet valve opens. The piston moves down. A mixture of petrol and air is sucked into the cylinder.

Compression stroke
The inlet valve closes. The piston moves up. The mixture is squeezed into a small space above the piston.

Power stroke
A spark from the plug makes the mixture explode. Hot gases from the explosion force the piston down the cylinder.

Exhaust stroke
The outlet valve opens. The piston moves up. The burnt gases (exhaust gases) are forced out of the cylinder.

Trucks

flat bed truck

truck with rubbish skip

tipper

tanker for bulk liquids, powder or grain

truck with drawbar trailer

articulated truck with container

Trucks (lorries) are large vehicles used for carrying goods by road. The heaviest can weigh 40 tonnes or more. Most are powered by diesel engines and can have as many as sixteen gears.

Trucks are often built with the same basic cab and chassis (frame), but different bodies are bolted on depending on the load to be carried. Special trucks can tip, carry liquids in a tank or keep food refrigerated.

Some trucks are in two parts. The front part, with the engine and cab, is called the tractor unit. This pulls the rear part, a trailer which carries the load. Trucks like this are known as articulated trucks, because the whole unit bends where the two parts join. As the parts can be separated, the trailer can be loaded before the tractor unit comes to collect it.

Most trucks are owned by haulage companies and hired out with their drivers. Some travel abroad or carry containers to and from the docks. Bigger trucks may have a bed, a basin and even a microwave oven so that the driver can live in the cab on long journeys.

Container Traffic

Many trucks carry their loads in containers. These are large sealed boxes which are made in standard sizes. The containers can be lifted straight on to railway trucks or cargo ships, or stored as they are in warehouses. This makes it much easier to handle different kinds of freight and saves a lot of money.

▶ An American articulated truck made by White's.

▼ Special trailers are made to carry very long loads. The longest load ever moved was a gas storage vessel, 83.8 m long.

TRANSPORT TECHNOLOGY

Coping with weight

Trucks can weigh up to several tonnes. They used to be very heavy to steer and difficult to stop in an emergency. Truck drivers had to be very strong. Modern trucks have power-assisted steering and brakes. Double wheels or broad tyres help spread the truck's weight to prevent it sinking in soft ground. For really soft ground special trucks with 'caterpillar' treads are used. These do not have wheels. Instead, they run on long metal belts that slide over rollers.

Flashback

Trucks, powered by steam or electricity, first appeared in the 1890s. Trucks with petrol engines were introduced in the early 1900s and were widely used during World War I (1914–1918) for carrying supplies. During the 1920s, reliable diesel engines became available. These were more long-lasting and economical than petrol engines and have been used on trucks ever since.

◀ A mammoth Caterpillar dump truck for use in quarries and open cast mines.

Longest truck
The Arctic Snow Train, originally built for the US Army, is 174 m long and has 54 wheels.

Very large trucks are sometimes called juggernauts. Juggernaut is a name of a Hindu temple and god whose image is traditionally carried in a huge, unstoppable chariot.

BULLDOZERS

These are tractors with a large blade in front which is used to level rough ground or to give it the right slope for building work. The tractor has caterpillar tracks so that it can travel over uneven or soft ground. The blade is usually square across the tractor so that the earth is pushed in front. Sometimes one end of the blade is slightly in front of the other so that the earth is pushed to one side. The blade can be raised or lowered hydraulically according to the amount of earth to be moved. Very small bulldozers are called calfdozers.

◀ This bulldozer weighs 93 tonnes and has a 770 horsepower engine. It is one of the largest tracked vehicles of its type in the world.

Farm machinery

1826 First practical reaper for cutting crops, designed by the Reverend Patrick Bell.

1902 First petrol-driven tractor, the Ivel.

Many of the jobs on the farm that used to be done by hand are now done by machines. These make the work far easier. Early machines were simple tools like spades and small ploughs to turn the soil, or scythes to cut corn (wheat). Today's more complicated farm machines can do almost everything from milking cows to spraying weeds.

▶ A plough pulled by a tractor. The coulter cuts into the soil, the share digs it up and the mouldboard turns it over.

mouldboard share coulter

▶ A combine harvester cuts the corn and separates the grain from the straw.

Saving time and effort

One of the most important machines is the tractor. This is a powerful vehicle with large wheels. Tractors are used to pull machines and carry heavy loads.

Farmers prepare their soil carefully before they plant their seeds. First, they turn over the soil with ploughs. This buries the weeds. Then they break up the big lumps of earth left by the plough with cultivators or harrows. A seed drill plants the seeds in long, straight rows and covers them with soil at the same time. Other machines spread fertilizer or insecticide on the fields. All these machines are pulled by a tractor.

Special machines have been made to harvest different types of crop, such as hops, potatoes and peas. The combine harvester cuts the corn (wheat) and separates it from the stalks. Other machines dry corn in the barn after harvest and grind it into meal to feed the animals.

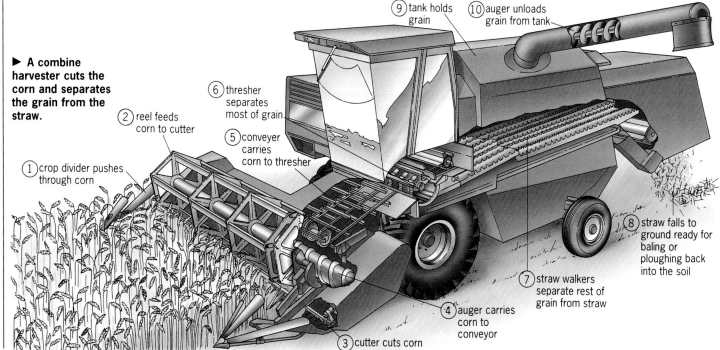

1. crop divider pushes through corn
2. reel feeds corn to cutter
3. cutter cuts corn
4. auger carries corn to conveyor
5. conveyer carries corn to thresher
6. thresher separates most of grain
7. straw walkers separate rest of grain from straw
8. straw falls to ground ready for baling or ploughing back into the soil
9. tank holds grain
10. auger unloads grain from tank

Bicycles

Modern bicycles range from the ultra-light racer to sturdy mountain bikes built for cross-country riding. But all bicycle frames are hollow, making a light, strong structure. The frame design is based on triangular shapes. Bicycle design is always developing, making use of new materials and ideas. Some cycles have tiny wheels and can be folded for easy storage.

A bicycle makes efficient use of your body strength to get you around. When you turn the pedals, the power of your legs is transmitted to the back wheel by a chain linking two gear-wheels. On many bicycles, you can make pedalling easier by changing gear. This either moves the chain so that it fits over gearwheels of a different size, or it moves tiny cogwheels inside the hub of the back wheel. For hill climbing, you use a low gear which gives plenty of force but not much speed. On the flat, a high gear gives less force but more speed. To slow down, you pull on the brake levers. This pushes rubber blocks against the wheel rims.

Flashback

The first bicycle, the hobby-horse, had no pedals, the rider pushing it along with his feet. Pedals were invented by a Scottish blacksmith who attached them to the back wheel by a lever. Then a French machine was developed with pedals attached to the front wheel. Wheels were made bigger and bigger as leg-power can turn a larger wheel fast; bikes with these huge front wheels were nicknamed penny-farthings (the names of large and small coins). Early cycles had iron wheels and later solid rubber ones which made them bumpy to ride. The safety bicycle, invented in 1885, had smaller wheels; the pedals were set on the frame and connected by a chain to the rear hub. After 1888 air-filled tyres made cycling much more comfortable, and soon after, bicycles were designed for women, with no crossbar to get in the way of skirts.

The hobby horse (velocipede) invented in Germany, 1817.

Kirkpatrick Macmillan's bicycle, 1839, with pedals.

The penny-farthing (1860s): one turn of the pedal made one turn of the wheel.

The 'safety' bicycle, 1885, with pedals and chain.

Sailing ships

The largest sailing ship ever built was the *France II* of 8,000 tonnes, launched in 1911.

The last big sailing ships were used until 1914 to carry fertilizer from South America to Europe around Cape Horn.

The area of sails of the *Preussen* was 5,574 sq m.

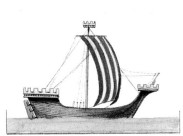

Cog (13th century). The sturdy wooden cog was developed in northern Europe and used as a trader and as a warship. Early vessels were steered with a large oar at the back. Later ships, like this one, had rudders.

Galleon (16th century). Galleons were used by both the British and the Spanish fleets at the time of the Armada. They were also used to carry gold and other booty from the Americas back to Europe.

Dhow (20th century). The Arabian dhow was developed many centuries ago and is still used for fishing and coastal trading today. The sail arrangement is called lateen rigging.

Junk (20th century). Junks have been used in China and the Far East for many hundreds of years. The fore-and-aft rigged sails have wooden spars to support and reinforce them.

Three-masted barque of the 1890s. Barques have fore-and-aft rigging on their mizen (rear) mast. Some were built of steel. Some are still used today as sail-training ships.

Six-masted schooner of the early 1900s. Schooners have fore-and-aft rigging on all their masts. The use of steel for masts and cables made it possible to build huge schooners with large sails.

Before the days of steam and diesel engines, merchant ships and warships had to rely on sails and the power of the wind to move them about. The earliest sailing ships had just one mast and sail. But as ships grew larger, more masts were added, masts became taller and several sails were carried on each mast.

On a sailing ship, the ropes and spars which support the masts and sails are called the rigging. Some ships had sails which lay along the direction of their length; they were fore-and-aft rigged like modern yachts. However, most large sailing ships had sails which lay across the direction of their length; they were square-rigged. The square-rigged clippers were the fastest sailing ships of all. Their name came from the way they could 'clip' time off their sailing schedules. In the 1850s, a clipper could carry a cargo of wool from Australia to Britain in just over two months.

Small sailing ships are still used for trading in many parts of the world. For example,

▶ This Japanese tanker has sails as well as an engine. It was launched in September 1980. The use of wind power can make substantial savings in the fuel needed to power the engines.

TRANSPORT TECHNOLOGY

dhows are used in the Middle East and junks in the Far East. Large, square-rigged sailing ships no longer carry cargo but some still survive in museums and for training young people at sea.

The first steamships carried sails to assist their engines. Today, ship designers are starting to look at this idea again as they search for ways of saving fuel and cutting pollution from engines.

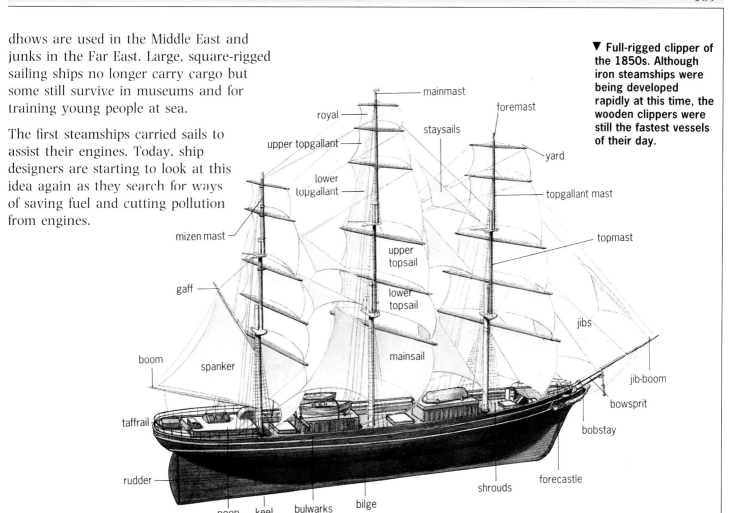

▼ Full-rigged clipper of the 1850s. Although iron steamships were being developed rapidly at this time, the wooden clippers were still the fastest vessels of their day.

SAILING

Sailing boats use the force of the wind on large sheets called sails to push them along. Sails used to be made of canvas but synthetic materials such as nylon are now normally used. To move in a particular direction the line of the sails is changed so that the wind blows across them. A sail acts rather like the wing of an aircraft. As the air flows across it, pressure builds up on one side. By angling the sail, the sailor can use the sideways pressure to push the boat forward. The keel (centreboard) stops the boat from moving sideways. The boat can sail with the wind behind it, to the side of it, or even slightly ahead of it. By tacking (zigzagging), it can move against the direction of the wind.

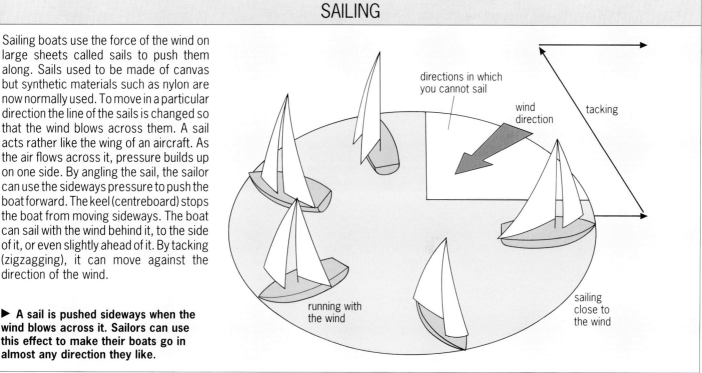

▶ A sail is pushed sideways when the wind blows across it. Sailors can use this effect to make their boats go in almost any direction they like.

Ships

First modern ship
The *Great Britain*, designed by Isambard Kingdom Brunel and launched in 1843, was built of iron and had a screw-propeller.

World's largest ship
The oil tanker *Seawise Giant* (564,000 tonnes) was 458 m long and 69 m wide. She was completed in 1976, but destroyed in a rocket attack in 1988.

Stopping distance
Oil tankers are so massive that it can take them over 10 km (6 miles) to stop even with their engines in full reverse.

▶ Ships carry passengers, cargo, vehicles and aircraft.

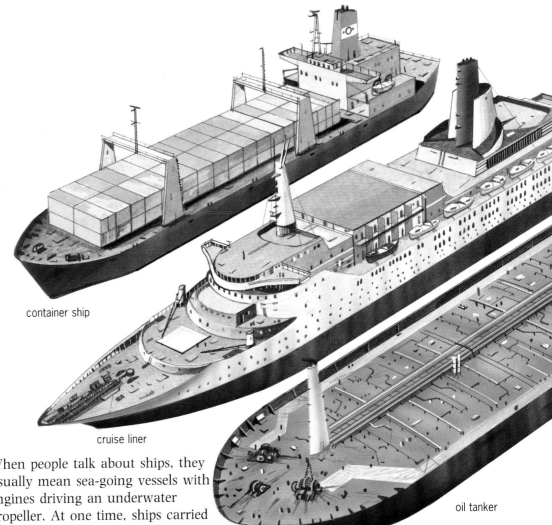

container ship
cruise liner
oil tanker

When people talk about ships, they usually mean sea-going vessels with engines driving an underwater propeller. At one time, ships carried sails and relied on the wind, but there are very few large sailing ships in use today.

Most ships have one propeller (also called a screw), but some have two or even three. Usually, the propeller is driven by a diesel engine, but a gas turbine or a steam turbine may be used instead. With a steam turbine, the steam comes from a boiler which uses the heat from burning oil or even from a nuclear reactor.

Cargo ships

At one time, a cargo ship would carry many different cargoes depending on what needed to be transported from port to port. Today, most cargo ships are specially designed to carry one type of goods only. The cargo might be several hundred cars, containers packed with washing machines, or grain pumped aboard through a pipe.

Passenger ships

Nowadays, passenger ships are either ferries, or luxury liners taking people on holiday cruises. However, before air travel became popular, huge passenger liners provided regular services between all the major ports of the world. When people had to travel long distances around the world, they went by sea.

TRANSPORT TECHNOLOGY

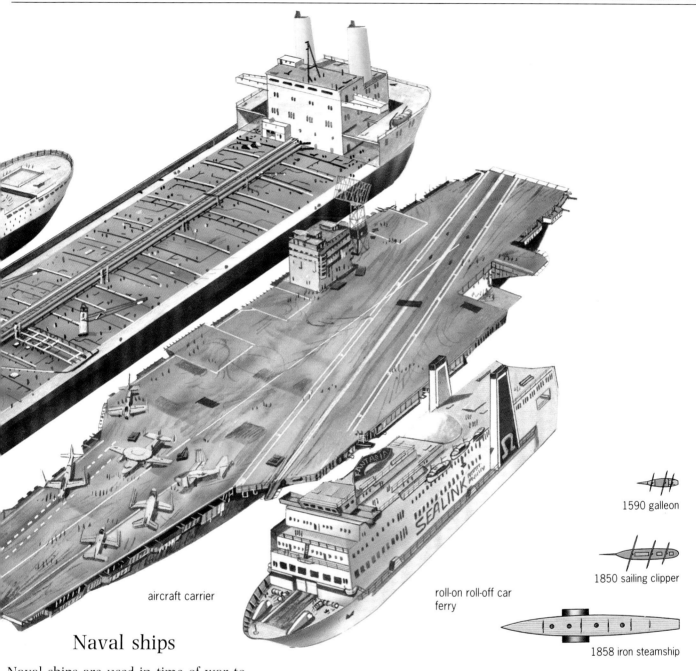

aircraft carrier

roll-on roll-off car ferry

Naval ships

Naval ships are used in time of war to hunt and destroy enemy ships, submarines and aircraft and to launch missiles against targets on land. Aircraft carriers are the largest naval vessels of all; the biggest can carry over 90 aircraft. Frigates and destroyers are used for escorting and protecting other vessels. Minesweepers hunt for explosive mines which might sink other ships. They have plastic or aluminium hulls which do not set off magnetic mines. Most naval vessels are packed with radar and other electronic equipment which can detect and track missiles or torpedoes launched against them.

▶ Ships have become larger over the years. The largest ships of all were the giant oil tankers built in the 1970s.

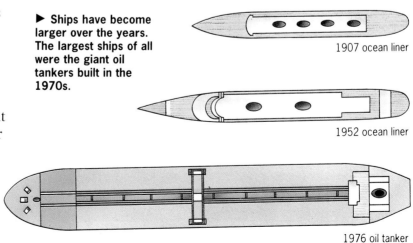

1590 galleon

1850 sailing clipper

1858 iron steamship

1907 ocean liner

1952 ocean liner

1976 oil tanker

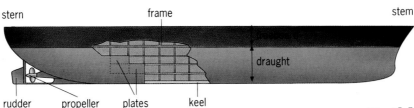

▶ The hull of a ship.

ISAMBARD KINGDOM BRUNEL

Brunel was the son of another famous engineer, Sir Marc Isambard Brunel. As a boy, he showed great skill at drawing and geometry. After leaving school he worked for his father. In 1826, when he was only 20, he was put in charge of a project to build a tunnel under the River Thames in London. The job almost cost him his life. He nearly drowned when water flooded into the tunnel and swept him away.

In 1830, Brunel won a competition to design a bridge to span the Avon gorge at Clifton in Bristol. But work on the Clifton suspension bridge was delayed and it was not completed until after his death. The turning point of Brunel's career came in 1833 when he was made chief engineer of the Great Western Railway Company. His main task was to build a railway between London and Bristol. The project took eight years and was one of the finest engineering achievements of its day. Brunel's genius did not stop there. During his lifetime he designed other railways and tunnels, guns, armoured barges and even a prefabricated hospital for use in the Crimean War. He designed many bridges using revolutionary new techniques.

Brunel was also interested in steamships. He built a wooden paddle-steamer, the *Great Western*, launched in 1838. This was the first steamship to cross the Atlantic. His second ship, the *Great Britain*, was a revolution in design. Launched in 1843, it had a screw-propeller and was built of iron. At 3,300 tonnes, it was easily the largest ship of its time and was the forerunner of all modern ocean-going vessels. Shortly before he died, Brunel designed another steamship, the *Great Eastern*, a 19,000-tonne giant with room for 4,000 passengers.

▲ The *Great Britain* being launched in Bristol in 1843. She was the first ship of her kind to cross the Atlantic.

Building a ship

The basic parts of a ship have not changed very much over the years. However, modern ships are built of steel and other metals, and not of wood as the early ships were. Building starts with the keel, followed by the stem, stern and frame. The plates of the hull, deck and inside compartments are then added, together with the engines. The ship is normally built on a gently sloping slipway. When it is ready to be launched, it is allowed to slide slowly down the slipway into the water. It is then floated to a quay where the remaining parts are added. In a modern shipyard, the ship may be built in separate sections which are joined together on the slipway.

Large ships need to be very strong to prevent their long hulls bending and cracking in heavy seas. Many are fitted with stabilizers (tiny, movable underwater 'wings') to stop them rolling too much.

Flashback

Ships driven by steam-engines started to replace sailing ships at the beginning of the 19th century. At first, steamships were propelled by huge paddle wheels at the sides, but more efficient screw propellers were being used from about 1840 on. Steamships needed huge supplies of coal as fuel, so they became much bigger. Iron rather than wood became the main building material and later steel. By the end of the 19th century, the engine rooms of the biggest liners were almost as large as cathedrals.

From about 1900, steam turbines began to replace steam-engines in the largest ships. Steam turbines are still used in some large ships, but with a nuclear reactor or oil to heat the boiler, rather than coal. Today, most ships are propelled by diesel engines, though some fast naval vessels use gas turbines similar to those in aircraft.

Submarines

Submarines are ships which can travel under water as well as on the surface of the sea. Most are naval vessels, but some are used for engineering and exploration.

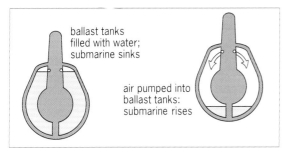

▲ A submarine can float or sink by emptying or filling its ballast tanks.

Diving and surfacing

Submarines have long hollow tanks on either side of the hull. These are the ballast tanks, which can be filled with air or water. To dive, the tanks are filled with water. Then the submarine is driven forward by its propeller, and the hydroplanes, which are like the tail on an aeroplane, force the submarine downwards. For the submarine to surface, the water in the ballast tanks has to be blown out by compressed air.

Submarines cannot use diesel and petrol engines when submerged, because these use up too much air and give out exhaust fumes. Instead, small submarines use electric motors and batteries. Larger ones are nuclear powered.

Submarines for exploration

These special submarines, or mini-subs, are of many different sizes and shapes. Within the hull there is a spherical cabin built of very thick metal. The crew look out through windows, and use powerful lights because there is no sunlight at great depths. These mini-subs are very useful for helping to position and repair oil rigs, exploring sunken wrecks and searching the sea-bed for valuable minerals. There are often movable arms attached to the hull which can hold tools or pick objects up. Divers cannot work at the enormous depths which these mini-subs can reach.

Flashback

The first vessel which could travel submerged or on the surface was built by a Dutchman who propelled it by oars from Westminster to Greenwich on the River Thames in 1620. Many people tried to improve on this idea until, in 1878, John Holland built his *Holland I* and *II*. The modern submarine is descended from these vessels. By 1914 there were nearly 400 submarines in existence. Submarines did not change much until the first nuclear submarine was built in 1954.

The first nuclear submarine, the USS *Nautilus*, was built in 1954.

In 1960, the submarine *Trieste* reached a record depth of 11,000 m (nearly 7 miles).

The world's biggest submarines are the Russian Typhoon Class. They weigh about 26,500 tonnes and are 170 m (558 ft) long.

▼ Mini-subs like this are used for exploration and oil rig maintenance. They can work at much greater depths than a diver.

Docks and ports

People have travelled the seas for thousands of years. Places on the coast where there is shelter from the open sea have always been good spots to start a port. All over the world natural harbours have grown up to be important ports and cities based on trading ships. The Middle Eastern port of Muscat, in Oman, has a fine natural harbour. The deep-water bay is protected by mountainous headlands which jut out from the jagged mountains that have protected the city from landward invasions. Muscat has been a trading port on the Gulf for over 500 years.

It was so successful that it remained an independent city state until it became the capital of modern Oman. In other places the natural landscape has been changed to make a better port. Rotterdam, in The Netherlands, was founded in the 13th century as a small fishing harbour 25 km (15 miles) up a river from the North Sea. Today Rotterdam is the world's busiest port. Ships dock all the way to the sea along a new channel dug at the end of the last century.

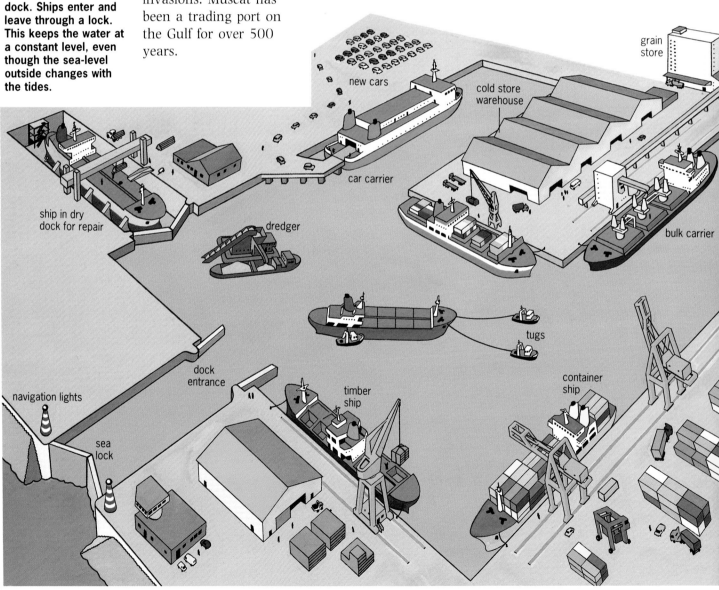

▼ Ships being loaded and unloaded inside a dock. Ships enter and leave through a lock. This keeps the water at a constant level, even though the sea-level outside changes with the tides.

◀ Containers are loaded onto ships in Rotterdam in The Netherlands. Rotterdam is called an *entrepôt* because goods are imported, then re-exported to another country such as Germany.

Fishing ports

Some ports just handle fish. There are many small fishing ports along Portugal's coast. The trawlers catch sardines, squid, cod and sole in the Atlantic Ocean. Every day you can see men and women sitting on the harbour-side surrounded by piles of nets and floats. There is always a lot of work to do mending holes in the nets and packing the fish in boxes to be sold. Every evening the narrow streets of Peniche, a small fishing port north of Lisbon, smell of barbecued fish. Outside each house a few sardines are slowly grilled over the charcoal, in a small round container.

Aberdeen is a fishing port on the eastern coast of Scotland. One of the most important fish landed in Aberdeen's port is herring. While most fish caught at sea are sold fresh, herring are usually preserved and then sold. Over more than 200 years a big industry has grown up in Aberdeen to preserve herring. The fish are gutted and then put in salt water in wooden barrels.

Passenger ports

Dover is a passenger port in the south-east of England. It has been a port since Roman times, as it is just 34 km (21 miles) from the French port of Calais. Dover is an important port for ferries carrying people across the English Channel to the European continent. The port sits below Dover's famous white chalk cliffs. The ferry passengers show their tickets and passports in a terminal building and walk onto the ship.

The ferries also carry cars and lorries. Drivers wait in large car parks to drive onto the ships. This sort of ferry is called 'roll-on roll-off', because that is just what the motor vehicles do. For the lorries carrying goods this makes the movement of cargo much faster. It saves the time spent unloading and loading from lorry to ship to lorry again. At the busiest times a ferry comes and goes every half-hour to and from the French ports of Calais and Boulogne and to Belgium and The Netherlands.

Largest port
New York and New Jersey, which has a navigable waterfront of 1,215 km (755 miles) stretching over an area of 238 sq km (92 sq miles).

Busiest port
Rotterdam, which handled 272·8 million tonnes of sea-going cargo in 1988. Rotterdam is also the largest artificial harbour, and covers an area of 100 sq km (38 sq miles).

Canals

Longest canal
The Volga to Baltic canal. Length: 2,977 km (1,850 miles) of rivers and linking canals.

▶ **The Corinth canal cuts a passage through Greece 6·3 km (3·9 miles) long from the Gulf of Corinth to the Aegean Sea. A passenger ship is pulled by two tugs.**

▼ **A fully loaded container ship passes through the Miraflores lock of the Panama Canal.**

A canal is an artificial river. Most have been built for boats, although others are used for irrigation or drainage. People have always used rivers as a cheap and easy way of moving heavy loads. Where there was no river, or where rivers flow too fast or are too narrow for boats, canals were built. The first canals were built in Mesopotamia in about 4000 BC.

The canal boom

In Britain during the Industrial Revolution people needed to transport heavy loads of coal and other materials between mines and factories. From the 1780s onwards a whole network of canals was built. These linked the industrial areas of the Midlands and North with London. Talented engineers, including James Brindley, William Jessop and Thomas Telford, designed them, and Irish labourers, known as navvies, did much of the constructional work. During the 19th century a canal network was built in North America linking the main industrial centres. Over a hundred years ago ship canals were built to join seas and oceans. These waterways dramatically cut the time taken by ships to travel around the world.

Linking seas and oceans

The **Suez Canal** links the Red Sea and the Mediterranean. It was designed by the Frenchman, Ferdinand de Lesseps, and has no locks. It was completed in 1869, and from then on, ships travelling from Europe to India and the East no longer had to sail around Africa. From 1967 to 1975 the Suez Canal was blockaded because of the Arab–Israeli Wars and could not be used.

In 1895 the **Kiel Canal** opened, making the journey from the Baltic to the North Sea much quicker and safer. Centuries before that, the Vikings had avoided this dangerous sea journey by dragging their ships overland on huge wooden rollers.

The **Panama Canal** was built to allow ships to go from the Atlantic and Caribbean to the Pacific without travelling around the continent of South America. It was finished in 1914, and its 82 km (51 miles) and six enormous locks contain more material than any other man-made structure in existence. At one point the canal rises 22 m (72 ft) above sea-level. During the many years of its construction, over 25,000 workers died from disease or accidents.

Linking rivers and lakes

The **St Lawrence Seaway**, opened in 1959, allows large ships to reach Lake Ontario from Lake Erie and so opens up 3,830 km (2,380 miles) of waterway, from the west of Lake Superior to the Atlantic. The New York State Barge Canal joins Lake Erie with the Hudson River.

The **Illinois Waterway** was built to connect with the Mississippi River, providing a north–south route from Chicago to Mexico. The Gulf Intracoastal Waterway links Mexico with Florida by a safe inland route.

In Europe a network of canals connects the Rhine, the Rhône and other main rivers. This means that large loads can be transported across national borders throughout the continent. New canals are being constructed in France and Germany.

LOCKS

A lock is used for raising or lowering boats and ships. In the early days of canal building, engineers often made the canal follow a constant height, called a contour. If the canal had to go over a hill rather than round it they had to build locks, which are like open boxes, containing water, with gates at each end. The lock is needed because the water outside the two gates is at different levels. By altering the water-level in the lock it is possible for a boat or ship to move from one water-level to another. A big hill needs a number of locks, as the greatest height a canal boat can be raised or lowered in one lock is about 9 m (30 ft).

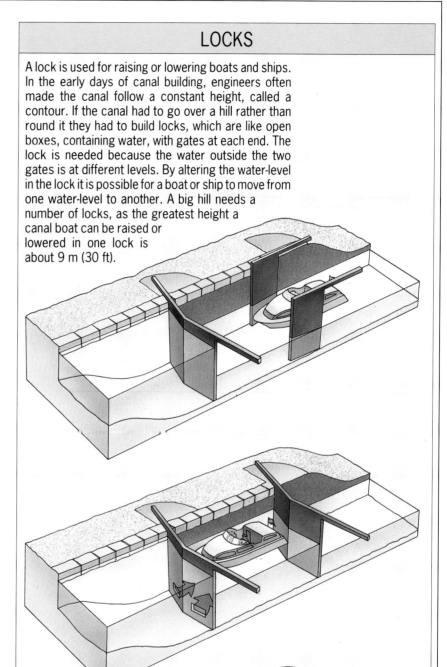

▲ How a boat passes through a pound lock. The boat enters when the lower gates are open. When all gates are shut, water is let into the lock through a sluice (small flap). When the water-level has risen, the upper gates are opened and the boat leaves the lock.

Bridges

Bridges can carry roads, railways or footways. Most are fixed but some can be raised or swung round. With all bridges, the problem for engineers is to design and build structures which will not sag or crack under the weight they have to carry. There are several ways of solving this problem.

Beam bridges have rigid beams which are supported at each end. The earliest bridges used this idea. They were just tree trunks or slabs of stone resting between the banks of a stream. In modern beam bridges, the beams are often long, hollow boxes made of steel or concrete. This makes them light but very strong. Bridges constructed like this are called **box girder** bridges.

▲ **The Europa bridge on the Brenner autobahn between Austria and Italy. It is a box girder bridge, which is a modern development of the simple beam bridge.**

▼ **Truss girder bridge**
A type of beam bridge which uses a rigid steel framework as a beam.

▼ **Arch bridge**
The weight is supported by one or more curved arches. The biggest steel arch bridge in the world is across the New River Gorge in West Virginia, USA. It is 518 m long.

▼ **Clapper bridge**
A type of beam bridge in which two or more beams are supported by piers.

▼ **Simple beam bridge**
A single beam rests across the river.

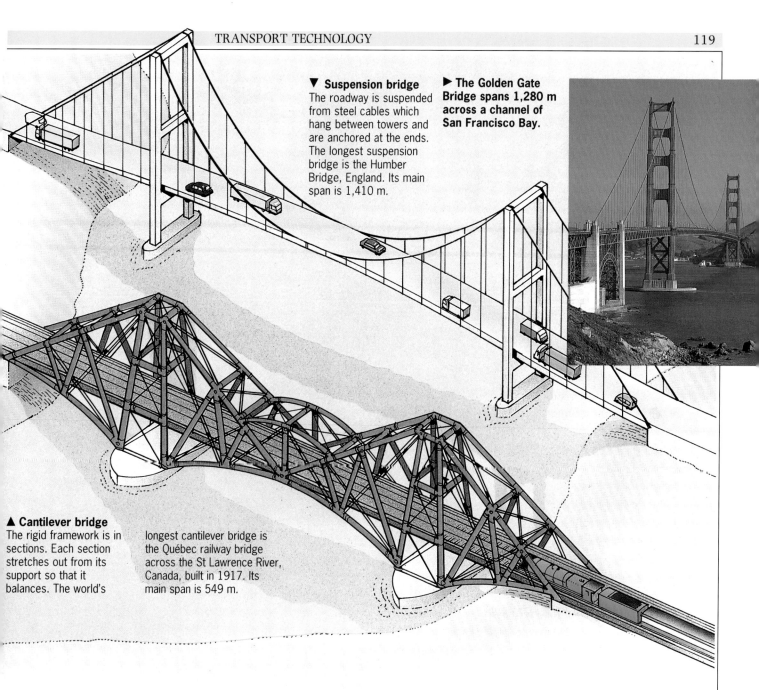

Suspension bridge The roadway is suspended from steel cables which hang between towers and are anchored at the ends. The longest suspension bridge is the Humber Bridge, England. Its main span is 1,410 m.

▶ The Golden Gate Bridge spans 1,280 m across a channel of San Francisco Bay.

▲ **Cantilever bridge** The rigid framework is in sections. Each section stretches out from its support so that it balances. The world's longest cantilever bridge is the Québec railway bridge across the St Lawrence River, Canada, built in 1917. Its main span is 549 m.

Cantilever bridges have long, rigid sections like beam bridges. However, each section is supported in the middle rather than at its end.

Arch bridges take the strain of the main span with an arch that pushes on the ground at each end. Modern arch bridges often have a light, open structure.

Suspension bridges are best for very large spans. The span is suspended from steel cables which hang between towers. To take the strain, the ends of these cables are anchored in the banks. Older suspension bridges sometimes have chains rather than cables.

▲ The Iron Bridge at Coalbrookdale, England, crossing the River Severn. It was the first cast iron bridge in the world, built by Abraham Darby and completed in 1779.

The oldest recorded bridge was built across the River Nile around 2650 BC. The clapper bridges of Dartmoor and Exmoor in England may be even older.

Roads

▼ This photograph shows the early stages of construction. The crushed rock will be off-loaded from the trucks and spread on the surface of the earth to form the lowest level of the foundation. In the background a bulldozer is levelling the surface.

Think about your recent journeys. You probably did most of them on roads. A road is usually the best route from one place to another. But it will not always be the most direct way. Sometimes natural obstacles like rivers or hills have to be avoided. Modern roads often bypass town centres or residential areas to avoid traffic jams or being a nuisance to residents. Country roads twist and turn round people's property.

Traffic needs to travel safely and efficiently on our roads. There are rules about which side of the road we use and how fast we can go. Lights, signs and road markings help control traffic. Direction signs help us find our way.

Types of roads

Motorways (highways or freeways in the USA) have several lanes in each direction separated by a central reservation. They have no sharp bends, steep hills, crossroads or roundabouts. Traffic can travel a long way at high speeds without stopping.

Other main roads are often dual carriageways. These also have a carriageway in each direction, but there are roundabouts and crossroads.

Urban ring roads provide routes around city centres to improve traffic flow. Residential areas are served by networks of smaller roads. In country areas narrow, winding roads often follow the routes of ancient tracks.

Road construction

Before a new road is built, the best route has to be chosen. Ground surveys and aerial photographs are used. The damage to the countryside, the inconvenience to people who live nearby, and the cost all have to be considered. Surveyors mark out the chosen route ready for the earth-moving vehicles to move in. These huge, expensive machines do the work of many people. A machine called a scraper loosens the top layer of soil, which is pushed aside by large bulldozers.

The road has to be as level as possible, so cuttings are made through hills and the soil used to fill in small valleys and form embankments. Bridges and tunnels may also be needed. The engineers then lay foundations of crushed rock so that the road can carry the weight of the predicted traffic. Sometimes cement or bitumen is added to the soil to form a solid base. Then a layer of concrete, called the base course, is put on the foundations. A machine called a spreader adds the top layer of asphalt or concrete. The concrete is often strengthened with steel mesh. Finally markings and signs showing the new road's number are added. The road can now be opened and shown on maps as ready for traffic.

◀ This *autobahn* in Germany is raised on concrete stilts high above the ground. The engineers have planned it so that it does the least possible damage to the farmland below and so that the hills are less steep for traffic.

Motorway (highway) speed limits

	km/h	mph
Australia	96	60
Belgium	120	74
Germany	130	81
Italy	140	87
Turkey	90	56
United Kingdom	113	70
USA	88	55

The history of roads

The first roads were narrow, twisting, bumpy tracks used by people and pack animals. Hard, smooth roads were needed once the wheel had been invented. The first paved roads were built in Mesopotamia (now Iraq) in about 2200 BC.

Two thousand years later the Romans built hard, straight roads all over Europe and North Africa for their soldiers. Roman roads were paved with stones. They sloped to either side to let rain-water drain away.

For hundreds of years roads in Europe were neglected. In the 18th century horse-drawn stage coaches needed better roads, so many new private roads were built. John McAdam invented a new kind of road surface. The ground was first drained. A layer of egg-sized stones was laid and covered in smaller stones. The pounding of horse-drawn coaches ground these together and the gaps between became filled with dust and dirt. These roads were no good for the rubber-tyred vehicles that appeared this century, so modern road surfaces of tar or concrete were developed.

In 18th century Britain many new roads were built privately by 'turnpike trusts'. They charged travellers a toll to use their section of road. There are still some private toll roads today.

◀ This huge intersection of freeways occupies a lot of land in the city of Los Angeles in California, USA. The whole city is cut through and ringed by a complex system of such freeways. The density of traffic has resulted in pollution caused by exhaust fumes.

Structures

The oldest free-standing structures are the megalithic temples at Mgarr and Skorba in Malta, which date from about 3250 BC.

The tallest self-supporting tower in the world is the CN Tower in Toronto, Canada, which is 553·34 m (1,815 ft 5 in) tall.

6·7 million cubic metres (246 million cubic feet) of earth and rock have been excavated to make way for the Channel Tunnel.

The Great Pyramid in Egypt was built over 4,500 years ago. It was 146·5 m (480 ft) tall. It contains 2,300,000 blocks of limestone weighing up to 15 tonnes each, and would have taken 4,000 men 30 years to build. To produce such a structure today would take 405 men 6 years and would cost $1·13 billion.

We are surrounded by structures – buildings tall and short, factory chimneys, bridges and tunnels, escalators, modern dams and old water-wheels, even fairground rides and helter-skelters.

Firm foundations

Engineers face special problems when they design very large structures. They must make sure the ground will support their weight. This means they have to dig good foundations, so the base of the structure is sunk deep into the ground.

Where there is room, the base of the structure or its foundations may fan out to spread the weight over a greater area. The Eiffel tower and most electricity pylons are wider at the base than at the top. This also makes them more stable.

Builders must also make sure the ground is suitable to build on. Some rocks become bendy under great pressure. Others swell when full of water, then shrink during a dry spell. This often causes buildings to subside (sink). If one part of the building subsides more than the other, the walls may crack and the building may fall down.

On soft ground it may be necessary to sink the foundations deep into harder rock below, or to build special 'floating' foundations. In some parts of the world, earthquake-proof buildings are built on concrete 'rafts' that literally float on the Earth's crust as the earthquake waves pass through the ground below.

Tall structures

A serious problem with very tall structures is the wind. Wind speeds get higher as you go up a tall building.

Many really tall buildings and towers are designed to bend. If they can bend a little with the wind, they are less likely to snap.

▲ The Eiffel Tower in Paris, built in 1887–89, is made of iron and weighs 7,340 tonnes. It stands 300·51 m tall, and has 1,792 steps. It may sway up to 12·7 cm in high winds.

If you look at the lamp posts lining a motorway in a strong wind, you will see them swaying.

The wind is less of a problem if it can blow straight through the structure. Many towers, fairground wheels and tall cranes are made of iron or steel scaffolding. Often the metal bars are arranged in triangles. This gives them extra strength. Tall buildings attract lightning. The world's tallest tower, the CN Tower in Toronto, Canada, is struck by lightning about 200 times a year. Tall buildings have lightning conductors, metal structures that conduct the current harmlessly to the ground.

Underwater structures

Large underwater structures have many of the same problems as tall buildings. They are very heavy, and need firm foundations. They are also exposed to waves and strong water currents, rather like underwater winds. During winter storms, they may face waves 25 m (82 ft) high.

There are two main designs for oil platforms. Close to shore are the semi-submersibles (rigs that are partly underwater). They are floating structures, and have large air tanks below the surface to keep them stable. Strong chains anchor them to the sea bed, so they can be moved from place to place.

Fixed oil platforms have long steel legs that go down deep into the sea bed. They have to carry a lot of weight. An oil platform will have heavy drilling equipment, a hostel for the workers, cranes for unloading supply boats, a helicopter pad and lifeboats. The world's heaviest oil platform, *Gullfaks C* in the North Sea, weighs 846 tonnes and stands 380 m (115 ft) above the sea bed.

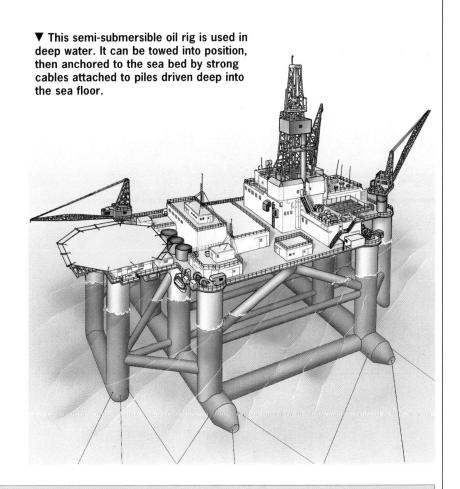

▼ This semi-submersible oil rig is used in deep water. It can be towed into position, then anchored to the sea bed by strong cables attached to piles driven deep into the sea floor.

THE CHANNEL TUNNEL

Tunnels are great time-savers for travellers along roads, railways and waterways. They can cut off bends, link up valleys and provide safe roads along the side of cliffs. They can also be used to hide roads and railways from view in areas of beautiful scenery.

Tunnelling through rocks is difficult. The tunnellers must use different drills to cut different types of rock. They must prop up the tunnel roof as they dig. They must line the tunnel to keep water from seeping in through the sides and flooding it. There must be enough air in the tunnel for people to breathe.

One of the greatest challenges has been the Channel Tunnel, which links England and France underneath the English Channel. The tunnel is 50 km (31 miles) long, and 37 km (23 miles) are under the sea. It will carry non-stop trains. There are actually 3 linked tunnels, one for trains running from England to France, another for France-to-England trains, and the third for services such as electricity, air conditioning and drains.

Safety is a great problem. Train passengers must be protected if fire breaks out in the tunnel, or if terrorists try to blow it up. Each tunnel can be sealed off completely, and there are lots of passages so that passengers can be led quickly into the service tunnel in an emergency. The tunnellers started working at each end of the tunnel, and finally met in the middle on 1 December 1990.

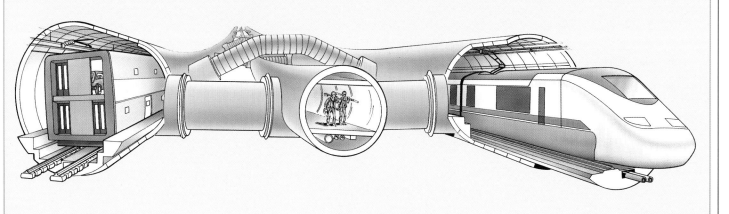

Cranes

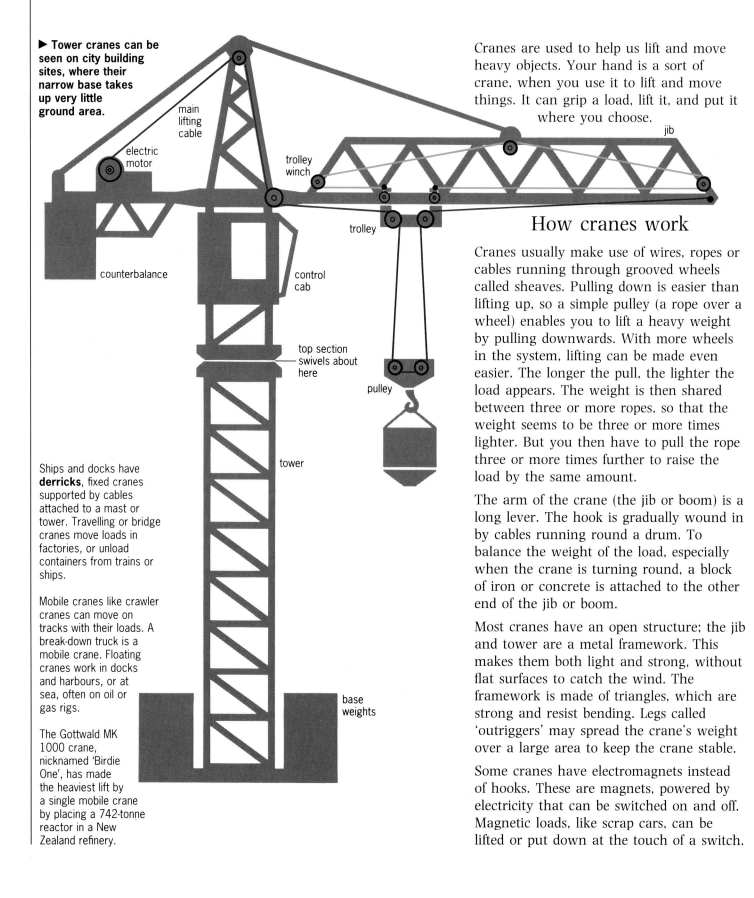

▶ Tower cranes can be seen on city building sites, where their narrow base takes up very little ground area.

Cranes are used to help us lift and move heavy objects. Your hand is a sort of crane, when you use it to lift and move things. It can grip a load, lift it, and put it where you choose.

How cranes work

Cranes usually make use of wires, ropes or cables running through grooved wheels called sheaves. Pulling down is easier than lifting up, so a simple pulley (a rope over a wheel) enables you to lift a heavy weight by pulling downwards. With more wheels in the system, lifting can be made even easier. The longer the pull, the lighter the load appears. The weight is then shared between three or more ropes, so that the weight seems to be three or more times lighter. But you then have to pull the rope three or more times further to raise the load by the same amount.

The arm of the crane (the jib or boom) is a long lever. The hook is gradually wound in by cables running round a drum. To balance the weight of the load, especially when the crane is turning round, a block of iron or concrete is attached to the other end of the jib or boom.

Most cranes have an open structure; the jib and tower are a metal framework. This makes them both light and strong, without flat surfaces to catch the wind. The framework is made of triangles, which are strong and resist bending. Legs called 'outriggers' may spread the crane's weight over a large area to keep the crane stable.

Some cranes have electromagnets instead of hooks. These are magnets, powered by electricity that can be switched on and off. Magnetic loads, like scrap cars, can be lifted or put down at the touch of a switch.

Ships and docks have **derricks**, fixed cranes supported by cables attached to a mast or tower. Travelling or bridge cranes move loads in factories, or unload containers from trains or ships.

Mobile cranes like crawler cranes can move on tracks with their loads. A break-down truck is a mobile crane. Floating cranes work in docks and harbours, or at sea, often on oil or gas rigs.

The Gottwald MK 1000 crane, nicknamed 'Birdie One', has made the heaviest lift by a single mobile crane by placing a 742-tonne reactor in a New Zealand refinery.

Printing

Printing is a way of making many identical copies from one original. For every page in a magazine, every stamp, poster, book, cereal packet or bus ticket there was one original. From the original (called camera-ready artwork) a printing plate was made from which hundreds, thousands and in some cases millions of copies can be printed.

Camera-ready artwork

Nearly all the printed material we see is made up of one or more of the following: type (which may be words or numbers), photographs and artists' illustrations. This book has all three. A bus ticket may have only type, while a stamp usually has type and a photograph or illustration.

A designer produces the camera-ready artwork. He or she asks an artist to prepare the illustrations. When they are complete, the designer gathers together the photographs and illustrations, and orders the typesetting.

Type is mostly set on a machine very similar to a word processor. All the different typefaces and sizes are stored electronically in the computer's memory. This page uses two different typefaces, one for the headings and marginal notes and another for the main text. Each typeface also appears in different sizes and in **bold** and *italics*.

When the typesetter has finished typing in the copy from the manuscript, the machine prints it out on special photographic paper or film. Each letter is formed by a laser beam moving across the paper and turning on and off, making the letters from a series of lines. The machine produces 100 lines in every centimetre. This might sound very slow but the machine can do this for as many as 1·8 million letters an hour.

The camera-ready artwork is produced by sticking the typesetting onto a piece of white card and marking the positions of photographs and illustrations. Today, this process can also be done on computers, which is how this page was produced.

Making the printing surface

Before any printing can be done, the words and pictures on the camera-ready artwork have to be transferred to the plate (printing surface), which is usually metal. This is done by putting a negative photograph of the original on a printing plate which has been coated with a light-sensitive material, and then exposing them to a bright light. When the coating is removed from the plate, the image of the original remains in those areas where the light has shone through the negative.

Type is solid black, but photographs and illustrations need a range of different tones from black to very light greys. The tones are produced by breaking the picture up

▼ Modern typesetting machines use lasers to set type. This is a letter 'a' which has been enlarged to show how it has been made from a series of lines.

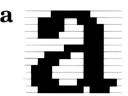

TYPE

Type is the term used to describe letters, numbers or any character used in printing. Originally each letter was cut out of a piece of wood or metal, but today most type is produced using a typesetting machine which is very similar to a word processor.

There are hundreds of different typefaces, but most fit into three categories; serif, sans serif (without serifs) and script.

   serif   sans serif script
typo gra *phy*

Each typeface can be set in a range of different styles,

Roman   *Italic*
Bold   ***Bold italic***

and in different sizes. Type is measured in points; there are 72 points in 25·4 mm (1 inch).

This book has been set in two typefaces. The main text has been set in 11 pt Photina, a serif face. The marginal notes and captions have been set in News Gothic, a sans serif face. The main headings are set in italic in 56 pt. The margin notes are set in 8 pt.

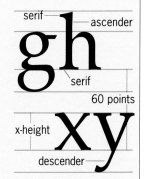

Lithography comes from the two Greek words: *lithos* meaning 'stone' and *graphia* meaning 'writing'.

Colour printing

Colour pictures are made up of tiny dots of magenta (a bluish red), yellow, cyan (a greenish blue) and black. All other colours can be produced from a combination of some or all of these colours.

Four different printing plates are needed. Each one makes its impression separately, one on top of the other. Black is always the last colour printed.

yellow plate

magenta plate

cyan plate

black plate

yellow

yellow + magenta

yellow + magenta + cyan

yellow + magenta + cyan + black

▲ A photograph (above) enlarged to show the dots of the halftone (below).

into a series of very small dots called a halftone. Each dot prints solid black, but little dots with big white spaces between them appear light grey and big black dots with very little white spaces between them appear almost black.

Printing

There are three main printing processes used today: letterpress, lithography and gravure.

Letterpress uses a printing surface with a raised image which is covered with ink and then pressed against the paper to transfer the image. The plates can either be flat for printing individual sheets of paper, or be curved around a cylinder, which is known as rotary letterpress. Rotary letterpress machines can print on a giant reel of paper (a 'web') at speeds of more than 500 metres per minute. Most newspapers are printed by letterpress.

Lithography This book was printed by lithography. It is the most widely used printing process today. The printing plate does not have a raised image. It uses the fact that grease attracts ink while water repels it. The printing image on the printing plate is greasy and so attracts the ink, while the non-printing parts are not greasy but wet, and therefore do not get inked.

Litho plates are usually made of aluminium, which can be easily wrapped around the plate cylinder. On the printing press, the plate comes into contact with two rollers, one for dampening it and the other for inking. The plate then comes into contact with the 'blanket' cylinder which is made of rubber. The inked image is offset (transferred) to the blanket cylinder and then on to the paper. Because the printing plate does not come into contact with the paper the process is frequently called offset litho. If single sheets of paper are fed into the printing press, the process is called sheet-fed offset. If a continuous roll of paper is used, the process is called web-offset. A machine using a roll of paper can print much faster than a sheet-fed machine.

Gravure In this process the printing surface, usually made of copper, has cells (like tiny wells) sunk into its surface. Each cell can be of differing depth and area. The whole plate is inked and then a metal blade scrapes all the ink off the surface leaving it only in the printing cells. Paper is then fed between the printing plate roller and another roller, and the high pressure between the two draws the ink out of the cells onto the paper. Making gravure plates is very expensive so this process is usually used for very long print runs such as magazines, packaging and wallpaper.

INFORMATION TECHNOLOGY

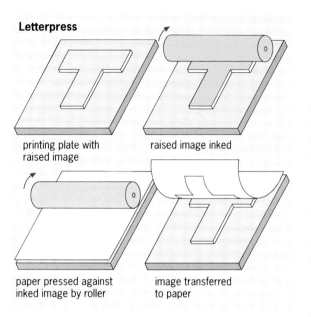

Letterpress
- printing plate with raised image
- raised image inked
- paper pressed against inked image by roller
- image transferred to paper

Photocopying

Photocopying is a quick and easy way of printing. In a photocopier, a bright light shines on the document you are copying. Lenses and mirrors project an image of the document, strip by strip, onto a rotating drum whose surface is charged with static electricity. The surface loses the charge under the light areas of the image but keeps the charge under the dark areas. A black powder called toner is attracted to the charged areas. It sticks to the drum, forming an image of the document in powder. As the drum rotates, it presses the powder against the next blank sheet of paper. Heat seals the powder onto the paper. In this way, the copy is printed.

The earliest surviving printed text is a Buddhist scroll printed in China between AD 704 and 751. Printing may have been invented more than 1,500 years ago.

◀▼ These pictures show the basic processes. Nowadays the printing plate itself is usually a cylindrical roller, or else the image is offset from a flat plate onto a cylinder base before it is printed.

Flashback

Printing was known in China, Japan and other parts of Asia many years before it came to be used in Western countries. The first books were made in China by a letterpress process, using hand-carved characters and designs on a flat block of wood.

By the 15th century, this process had spread along the trade routes of China and Europe. The Chinese had also invented paper, the ideal material for printing on. In about 1450, Johann Gutenburg of Mainz, Germany, developed movable metal type which could be used again and again. He cast individual letters and held them in place in a type mould. Then he adapted a wine press to print them. Lithography was introduced around 1800, and mechanical typesetting by 1900. Photo-typesetting systems, which produce type on film instead of metal, first appeared in the 1950s.

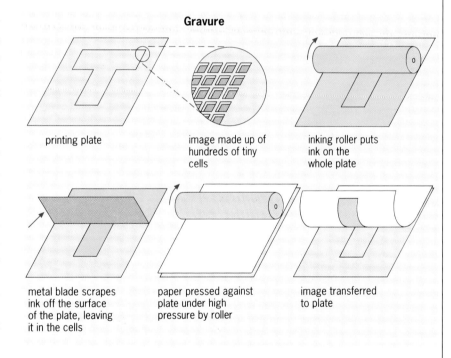

Gravure
- printing plate
- image made up of hundreds of tiny cells
- inking roller puts ink on the whole plate
- metal blade scrapes ink off the surface of the plate, leaving it in the cells
- paper pressed against plate under high pressure by roller
- image transferred to plate

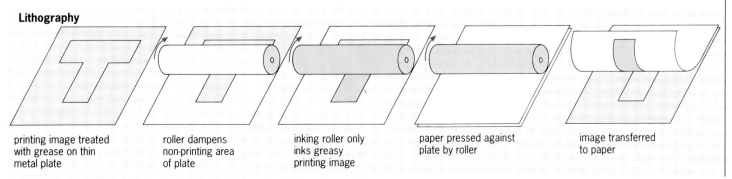

Lithography
- printing image treated with grease on thin metal plate
- roller dampens non-printing area of plate
- inking roller only inks greasy printing image
- paper pressed against plate by roller
- image transferred to paper

Paper

For every tonne of waste paper collected and reused, at least two trees are saved.

Paper is made from plant fibres matted together to form a sheet. For hundreds of years fibres were obtained from pulped cotton and linen rags, but during the last century it was discovered that paper could be produced from wood pulp. Now most of our wood pulp comes from conifers such as pines, spruces and firs.

When the logs have reached the pulp-mill, the bark is stripped off them. They may then be ground between heavy rollers or 'cooked' with chemicals to break the wood into fibres. These fibres are made into a thin slush, by adding water. The mixture then passes through 'beaters' which fray the fibres so that they will readily mat together.

The papermaking machine

The pulp of beaten and treated fibres next passes to the papermaking machine. When the slushy mixture arrives at the 'wet end' of the machine, it goes onto a fast moving belt of fine mesh. Some of the water drains away through the mesh, while still more is sucked off.

The remaining pulp, still about 80 per cent water, goes onto the rollers. These squeeze out even more water and press the fibres firmly together so that they form a sheet. The 'web' of paper is now strong enough to hold its own weight. It is led around a large number of heated rollers which continue to dry it. Finally the paper emerges in a huge roll.

There are hundreds of different types of paper, and many go through other processes. Some paper is moistened and passed through heated rollers which give it a glossy surface. Other papers are given coatings of china clay to make them into high-quality art and printing papers.

Recycling paper

Millions of trees are cut down each year to make paper. But it is possible to make perfectly good paper from waste paper. When waste paper is soaked in water, it breaks down into its original fibres. These can be used over and over again. Many newspapers are made of recycled paper which has been de-inked and cleaned.

▼ In a papermaking machine, liquid pulp is poured out onto a moving wire mesh. After pressing and drying, the fibres in the pulp form a continuous roll of paper.

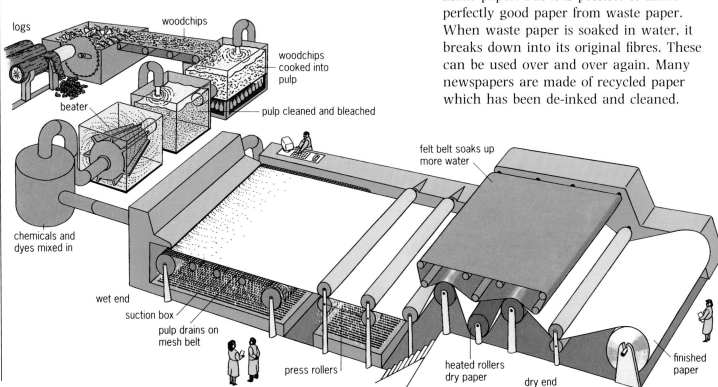

Cameras

A camera will make a picture of a person or a scene that you can keep, called a photograph. Its lens takes in the light from the person or scene and makes a small picture on the film inside the camera. The film stores the picture because it is covered with chemicals which are sensitive to light. If too much light falls on the film it will spoil the picture, so the film is kept in the dark until it is processed to produce the finished photographs.

Using a camera

A camera may need setting before you take a picture, though automatic cameras do the adjustments for you and simple cameras have nothing to adjust. The lens is set to take a clear picture of something close up or far away. The shutter is a door that opens for a fraction of a second to let in the light. The aperture is a hole that lets in the right amount of light. Both are set to let in more light on a dull day or less light on a bright sunny day. If there is not enough light, a flash will give extra light while you take the picture. When you take a photograph, you look through the viewfinder to see what the picture will look like. Then you press the button to open the shutter and take the picture. The film must then be wound on to get a fresh part in position.

Different kinds of cameras

Instant cameras can produce finished photographs a few minutes after they are taken. The film is in a flat package with everything needed to produce a picture. The package comes out of the camera and you can see the picture appear as you hold it in the light. Some special cameras, like TV and video cameras, do not use film at all; the pictures are made electrically. Others do not use light; X-ray cameras make pictures by sending X-rays through the body.

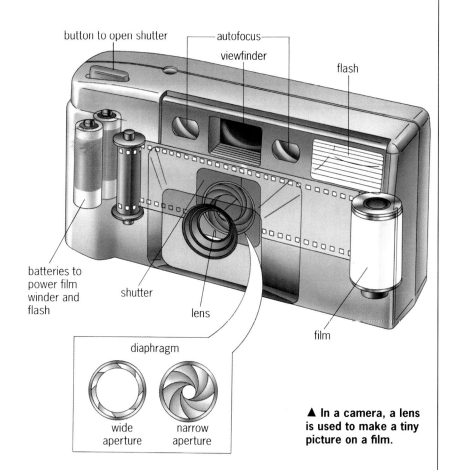

▲ In a camera, a lens is used to make a tiny picture on a film.

Flashback

In the 16th century the 'camera obscura' was invented in Italy. It was just a dark box with no windows and a small hole in one wall. It made a picture of the scene outside, but it did not make a picture that lasted. In 1826 a Frenchman, Joseph Niepce, made the first real camera. It was a wooden box with a lens at the front and it made the first permanent picture on a metal plate. The French inventor, Daguerre, and the British photographer/inventor, Fox Talbot, made improvements in photography. Then in 1888 an American, George Eastman, produced the first popular camera. This had a roll of film that took 100 pictures. Edwin Land (also American) invented a camera that could produce instant photographs in 1947.

Photography

▶ Early photographs, like this one taken in the 1880s by Frank Meadow Sutcliffe in Whitby, Yorkshire, had to be posed. This meant the people being photographed had to stay still for several seconds. It took that long for the light reflected off the subjects to affect the chemicals on wet plates in the camera. These plates were used before film was invented.

Photography is the art of taking pictures with a camera. Modern cameras adjust automatically to focus clearly on the subject and to let in the right amount of light. The film reacts to light so quickly that it can capture a sharp image even if the subjects are moving. The main skill is to compose the picture so that it is pleasing and shows what you wanted.

Posed photographs are carefully arranged to achieve the right effect. Candid shots are unposed, but the photographer can select the right angle and the right distance to get balanced pictures. Professional photographers become expert at this, but they also take many pictures of the same scene and then pick the one that turns out best. They may even have a camera with a motordrive that automatically takes many

▶ This colour photograph of a skier was taken using a sophisticated camera that can operate at speeds of less than one thousandth of a second. This means even the fastest action can be captured as a perfect image in sharp focus.

pictures one after the other when they press a button. They can later select the picture that shows the action at just the right moment.

Indoor photography

Indoor photography needs more skill, to get the lighting right. Modern flash cameras can provide enough light, but the flash may make the subject of the photo look shiny, or it may make people blink! Professionals in studios have special lamps with shades to give them good light without glare. There are also special fast films for taking pictures indoors without flash. To take photos in dim light, the shutter must be kept open a long time, so you should use a tripod to keep the camera steady. A tripod is also useful when you are using heavy lenses.

Special effects

There are many special effects that can be used with photographs. For example, paint may be blown on with an airbrush to change parts of a picture. A photograph can be 'cropped' so that parts of it are not shown when it is printed. The people in the foreground of one picture may be put on the background from another picture. Some of the photographs that you see in magazines never actually happened: they were put together.

Flashback

The first photograph was taken by a French inventor, Joseph Nicéphore Niepce, in 1826. He used a metal photographic plate which needed to be exposed for eight hours. In the 1830s Louis Daguerre developed this process and invented the first practical camera. At the same time an English scientist, William Henry Fox Talbot, produced a light sensitive paper which made a negative, from which positive prints could be made. It is this process which is used to make most photos today.

GUIDELINES FOR TAKING BETTER PHOTOGRAPHS

- Hold the camera steady and squeeze the shutter button gently.
- Do not have your fingers or camera case in front of the lens.
- Have the sun or light source to one side or behind you, and remember that bright sunlight creates deep shadow.
- Frame your picture carefully and do not cut off heads or feet.
- Think about the arrangement or composition of your shot. Would it be better with the camera upright or horizontal?
- Stand close to your subject and fill the frame, or the background will dominate the picture.
- Avoid unwanted background, and check you have not got objects like trees 'growing' out of someone's head.
- Do not always take people facing the camera. A sideview (profile shot) can be interesting.
- Do not have people looking straight at the camera when using the flash, it causes 'red eye'.
- Try to catch your subject unawares. Candid shots often look more natural.
- Arrange groups on different levels for added interest.
- Photograph from a low or high angle for an unusual result.
- To get a sharp picture of a moving subject follow it with your camera. This is called panning.

Radio

Frequency
200 kHz means that the transmitter is sending 200,000 radio waves every second.
100 MHz means that the transmitter is sending 100 million radio waves every second.

Speed of radio waves
299,000 km per second (186,000 miles per second). A radio wave could travel right round the world in the time it takes you to blink.

▶ How sounds are sent from one place to another by AM radio.

With a radio, you can listen to someone speaking into a microphone hundreds or even thousands of miles away. The sounds do not travel that far. Instead, they control the flow of radio waves from a transmitter. Radio receivers pick up the radio waves and use them to produce a copy of the original sound.

Sending sounds by radio

When someone speaks into a microphone, their voice sends vibrations through the air. The microphone turns the vibrations into tiny, changing currents called electrical signals.

A transmitter makes radio waves by making a current surge rapidly to and fro in an aerial. The waves are sent out in a continuous stream called a carrier wave. In the simplest type of transmitter, the signals from the microphone control the strength of the radio waves being sent out. This means that the waves pulsate to match the sound vibrations. Controlling radio waves like this is called amplitude modulation (AM).

The pulsating radio waves (called radio signals) are picked up by an aerial in the receiver. The receiver turns the pulsations into electrical signals which go to a loudspeaker. The loudspeaker makes vibrations in the air just like the ones that went into the microphone, so you hear a copy of the original sound.

Frequency and wavelength

Transmitters send out many thousands, or even millions, of

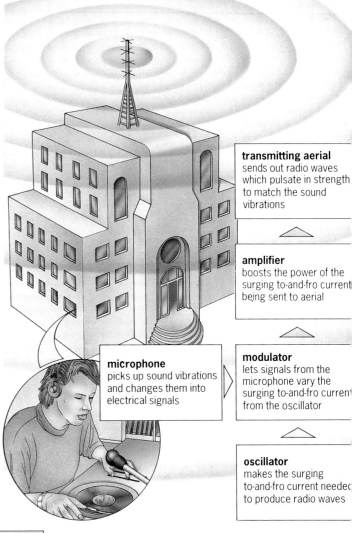

transmitting aerial sends out radio waves which pulsate in strength to match the sound vibrations

amplifier boosts the power of the surging to-and-fro current being sent to aerial

modulator lets signals from the microphone vary the surging to-and-fro current from the oscillator

oscillator makes the surging to-and-fro current needed to produce radio waves

microphone picks up sound vibrations and changes them into electrical signals

▼ Types of radio waves and their uses.

Type of radio wave	long wave	medium wave	short wave	VHF (very high frequency)	UHF (ultra high frequency)
used for	national broadcasting; AM radio; long distance ship communication	national broadcasting	international broadcasting	national broadcasting; high quality two-way radios; car phones	TV transmission
typical frequency	200 kHz	1 MHz	10 MHz	100 MHz	1,000 MHz
typical wavelength	1,500 m	300 m	30 m	3 m	0.33 m

▼ Radio waves can travel thousands of kilometres round the Earth, bouncing between the ground and the ionosphere.

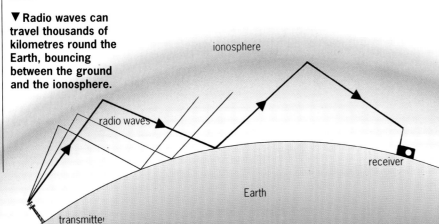

INFORMATION TECHNOLOGY

radio waves every second. The number of waves every second is called the frequency. It is marked on the tuning scale of a radio either in kHz (for kilohertz, meaning 'thousand waves every second') or MHz (for megahertz, meaning 'million waves every second'). Different stations use different frequencies, so you have to tune the receiver to select the one you want. The more radio waves are sent out every second, the closer they are together. The distance from one wave to the next is called the wavelength.

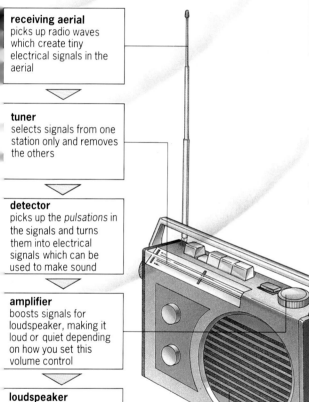

▲ Assembling radio sets in 1945, when peace was restored after World War II. Many families would spend the evening listening to the radio during the 1930s and 1940s.

receiving aerial picks up radio waves which create tiny electrical signals in the aerial

tuner selects signals from one station only and removes the others

detector picks up the *pulsations* in the signals and turns them into electrical signals which can be used to make sound

amplifier boosts signals for loudspeaker, making it loud or quiet depending on how you set this volume control

loudspeaker changes electrical signals into sound vibrations just like the ones that went into the microphone

main telephone network by radio. Ships and aircraft use radio for communication and for navigation; they can work out their position using signals from radio beacons. Television uses radio waves for transmitting pictures and sound. Spacecraft can be controlled by radio. So can model cars, boats and planes.

FM radio

VHF (very high frequency) radios use frequency modulation (FM) rather than AM. With FM, the signals from the microphone vary the frequency of the radio waves rather than their strength. FM is used for high quality stereo broadcasting, because it is less affected by interference than AM.

Flashback

Using radio waves

As well as sound broadcasting, radio waves are used for many other types of communication. Police, fire, taxi and ambulance crews use two-way radios for communicating with headquarters or with each other. Car phones are linked to the

The first radio sets used receivers with tiny crystals as detectors. Sound levels were so low that people had to use headphones. Later, receivers had amplifiers with large valves to boost the signals so that the sound could be heard through a loudspeaker. In the 1950s tiny transistors started to replace valves, and the first small, portable radios began to appear.

1895 Radio signals transmitted over a mile by Marconi
1901 Marconi picks up first radio signals transmitted across the Atlantic
1906 First broadcast of speech and music
1915 First transmission of speech across the Atlantic
1918 First 'superhet' radio receiver with tuning and reception like those of a modern set
1920 First regular public broadcasting (America)
1954 First transistor radio
1961 First stereo broadcast

Television

Television (TV for short) is a way of sending pictures so that events happening in one place can be seen somewhere else. A TV camera changes the pictures into signals. The signals can be transmitted (sent) using radio waves in much the same way as sounds are sent by radio. When a TV set receives the signals, it uses them to make pictures on its screen.

Many small, portable TV sets give black-and-white pictures. The main part of the set is the picture tube. At the narrow end of the tube there is a 'gun' which shoots out a beam of electrical particles called electrons. At the wide end there is a screen with a phosphor coating on the inside. The phosphor glows where the electron beam strikes it. This produces a spot of light.

To make a picture, the electron beam is pulled by electromagnets so that it zigzags its way down the screen. At the same time, the strength of the beam is varied so that the spot glows brighter or dimmer. The result is a picture made up of hundreds of horizontal lines. Along each line, the white bits are where the beam is at full strength. The 'black' bits are where the beam is so weak that the screen does not glow at all. When the picture is changed it is not changed all at once but bit by bit. Making the electron beam zigzag over the screen is called scanning.

◀ The picture on a black-and-white TV screen is made up of hundreds of horizontal lines which change between light and dark along their length. The lines are made by a beam of electrons which strikes the back of the screen and makes it glow.

1 electron gun shoots out a beam of electrons
2 magnetic coils bend electron beam
3 electron beam zig-zags down screen
4 screen glows where electron beam strikes it
5 brightness of glow is varied by changing strength of electron beam
fluorescent screen
moving spot

Making the pictures

The 'moving' pictures on a TV screen are not really moving at all. What you actually see is lots of still pictures flashing on the screen one after another. This happens so quickly that it looks like continuous motion. Films in the cinema use the same effect.

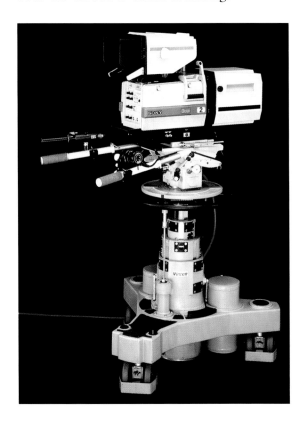

▶ A television studio camera. The operator can see what is being filmed in a small TV screen on top of the camera.

INFORMATION TECHNOLOGY

HOW A COLOUR PICTURE IS MADE

red picture
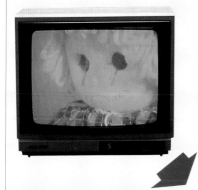

green picture

blue picture

pictures combine to give full colour picture

Your eyes can be made to see any colour, including white, by mixing red, green and blue light together in the right proportions. Colour TV uses this idea. A colour TV set gives you separate red, green and blue pictures on top of each other on the same screen. When you see the three pictures together, they look like a single picture in full colour.

Before signals can be transmitted, a colour TV camera has to split each picture into separate red, green and blue pictures. It does this using mirrors, filters and three different light-sensitive plates.

A colour TV set uses three electron beams to produce the red, green and blue pictures. The screen is covered with thousands of tiny phosphor strips which glow red or green or blue when a beam strikes them. Behind the screen there is a special grid called a shadow mask. This makes sure that only one of the beams can strike the red strips, one beam the green strips and one the blue. If you look at a colour TV screen through a magnifying glass, you can see the red, green and blue strips grouped in threes all over the screen. See if you can work out which strips are glowing if the screen looks white when you stand back from it.

◀ A colour TV picture is really three separate pictures, one on top of the other. When you look at the screen, you see the three pictures as a single picture in full colour.

Closed-circuit television
The signals from the camera are sent straight to a TV set and not transmitted. Security cameras in shops use closed-circuit television.

The average British viewer spends more than 25 hours a week in front of a television set.

The largest TV screen in the world was built by Sony for an international exhibition in Tokyo. It measured 45 m by 24 m.

The tallest television transmission mast in the world is at Fargo, North Dakota, USA. At 628 m, it is just over twice the height of the Eiffel Tower.

In Britain, TV pictures have 625 lines. 25 complete pictures are put on the screen every second.

In the USA, TV pictures have 525 lines. 30 complete pictures are put on the screen every second.

About 3,100 million people watched the 1990 World Cup finals played in Italy.

Sending the pictures

A TV camera uses lenses to pick up an image, just like an ordinary camera. However, a TV camera does not have film in it. Instead each image falls on a light-sensitive plate. A beam of electrons scans the plate, just as in a picture tube. This makes the plate give off a tiny, varying current. So, line by line, the picture is changed into electrical signals.

The signals from a TV camera are transmitted in a similar way to radio signals, by modulating (varying) radio waves. UHF (ultra high frequency) radio waves are used for TV transmission. UHF waves cannot bend round hills or large buildings, so, for good reception, a TV set needs an aerial which points straight at the transmitter.

◀ A family watching colour television in 1954.

Telephones

The very first words spoken over a telephone were:
'Mr Watson, come here, I want to see you!'
They were spoken by Alexander Graham Bell, calling his assistant in the next room.

Telephone conversation takes about a quarter of a second to travel from Britain to America. Calls are sent via a satellite 36,000 km (22,000 miles) above the Atlantic Ocean.

▼ When you speak into a telephone, the sound is changed into electrical signals. In your friend's telephone, electrical signals are changed back into sound.

A telephone lets you talk to people who are far away. It is the most widespread means of long-distance communication and there are over 525 million telephones throughout the world.

How does a telephone work?

When you telephone a friend, you probably use a handset with a separate mouthpiece and earpiece. As soon as you lift the handset off the base, your telephone is automatically connected to the local exchange. The exchange sends a dialling tone which means you can dial or tap in the number you want. The dialling unit sends out a series of electrical signals. These correspond to the number you enter and the exchange automatically connects you to the telephone line of your friend. Your friend's telephone will then ring.

Uses of the telephone

Emergency services such as fire, police and ambulance can be rung using a special priority number. In Britain, the number is 999. Businesses use telephones to communicate with customers and suppliers, but they also use them to link staff who are working in different rooms. People use mobile phones to make and receive calls in cars and trains or when walking. The phones are linked to the exchange by a network of radio transmitters and receivers. Computers in different places use telephones to send data to each other. Each computer is connected to the telephone by a device called a modem. This sends computer signals down the telephone lines instead of ordinary conversation.

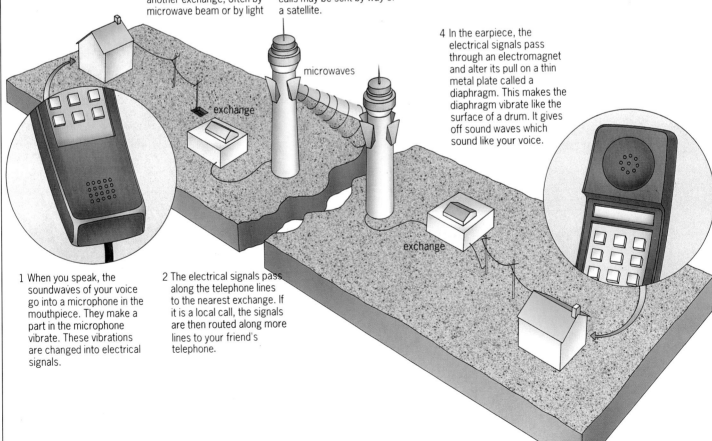

1 When you speak, the soundwaves of your voice go into a microphone in the mouthpiece. They make a part in the microphone vibrate. These vibrations are changed into electrical signals.

2 The electrical signals pass along the telephone lines to the nearest exchange. If it is a local call, the signals are then routed along more lines to your friend's telephone.

3 For a long distance call the signals are passed on to another exchange, often by microwave beam or by light pulses passing along an optical fibre. International calls may be sent by way of a satellite.

4 In the earpiece, the electrical signals pass through an electromagnet and alter its pull on a thin metal plate called a diaphragm. This makes the diaphragm vibrate like the surface of a drum. It gives off sound waves which sound like your voice.

INFORMATION TECHNOLOGY

TELEFAX

Telefax is the sending of text, photographs, maps, drawings and even handwriting across telephone lines. This process is called facsimile transmission.

The original page of text or pictures is scanned and the images are changed into electric current. The current is sent over normal telephone lines or by means of microwaves to a receiver. The receiver changes the electrical signals back into an image on a piece of paper. Some telefax machines can take text and pictures directly from one computer to another without having to put them on paper first.

Publishers and businesses use telefax to send information faster than it would travel in the post. Newspaper publishers use it to receive urgent news photos from overseas. Telefax can be used for more distant communications. Weather forecasters use it to receive photographs and other images produced by satellites orbiting the Earth.

Flashback

The first telephones were built in 1876 by the Scotsman Alexander Graham Bell in the USA. He demonstrated them by speaking to his assistant in the next room. In Bell's early designs, the earpiece was also the mouthpiece. The first telephone exchange was opened in 1878 in Connecticut, USA. It had just 21 customers. The connections were made by an operator who had to listen to the calls to find out when they had ended. Automatic telephone exchanges, without operators, were first built in the late 1890s. Customers could call each other up by pressing buttons or turning a dial on the telephone. The first transatlantic telephone service was set up in 1928. Conversations were carried by radio signals.

Bell was trained to follow in the footsteps of his family, who were all experts in teaching people to speak clearly. He went to the USA to continue this work and became convinced that he could teach totally deaf people to speak. But he was also interested in other kinds of science, and after he was successful in teaching two particular deaf students to speak, their fathers offered to help him with money for his other experiments. One of these experiments led to his invention of the telephone, and he set up a company to develop and make telephones for sale. This, his most famous invention, is remembered in the name of one of the great corporations of America, Bell Telephone Systems.

In 1898 Bell became President of the National Geographic Society, and was so convinced that one of the best ways of teaching was through pictures that he started the National Geographic magazine, now world famous for its superb colour pictures.

Telefax comes from the words telephone and facsimile. Facsimile comes from the Latin words *fax*, which means 'make', and *simile*, meaning 'like'.

Telefax was invented in 1843 by Alexander Bain, a Scottish clockmaker and inventor.

▼ **Bell demonstrated his telephone to a group of businessmen in 1892 by sending a call from New York to Chicago. He had built his first experimental telephone sixteen years earlier in 1876.**

Satellites

► An Intelsat communications satellite in geostationary orbit. This Satellite can carry up to 90,000 voice channels and three TV channels at the same time. A network of these satellites links 110 countries worldwide.

Telstar
Carried the first live television pictures across the Atlantic Ocean in 1962.

Weather satellites
The US government's National Oceanographic and Atmospheric Administration (NOAA) has weather satellites in polar orbits. These provide most of the cloud pictures seen on TV weather forecasts. Meteosat (European satellites) fly in a geostationary orbit above West Africa.

Communication satellites (comsats) and direct broadcasting satellites (DBs) are placed in a geostationary orbit 35,800 feet above the Earth.

The International Telecommunications Satellite Organization (Intelsat) is a body of over 110 countries which pay for a network of communications satellites in geostationary orbit above the Atlantic, Pacific and Indian oceans.

A satellite is something going in orbit round a larger object. The Moon is a natural satellite of the Earth. It travels round and round the Earth in a set path. The natural satellites of other planets are often called moons. There are also many artificial satellites in orbit round the Earth that have been put there by rockets.

A satellite's orbit is its path in space. The shape of the orbit is an ellipse - an oval shape. A satellite keeps going round a planet and does not fly off into space because the pull of gravity of the much bigger planet holds onto it. Weather satellites and communications satellites travel around the Earth in 24 hours, so they are always directly above the same point on Earth. This is called a 'geostationary orbit'.

What satellites are used for

Some Earth satellites are manned space stations where the crew can do scientific experiments and test the effects of weightlessness. Other space experiments are done by unmanned satellites. These satellites carry scientific instruments which are controlled from Earth by means of radio and other signals.

Unmanned satellites have many uses. Communications satellites relay telephone, television and radio signals between continents. Navigation satellites send out radio signals that help ships and aircraft work out their positions. Weather satellites photograph the cloud patterns for the forecasters. 'Spy' satellites take close-up photographs of the ground for military use. Astronomers use satellites to study space without interference from Earth's atmosphere.

Energy for satellites

Many artificial satellites have solar panels which absorb the sun's rays and use the energy to work their instruments. Some use nuclear power, but this is not often used because if the satellite breaks up the radioactive material could cause problems for other satellites. Satellites may have extra rockets on board which can be fired to change their orbit.

Radar

Radar (RAdio Detection And Ranging) can tell us the position of a moving or stationary object even if it is too far away to see or if it is dark or foggy. It is essential for air traffic controllers, who need to know the height and position of aircraft around busy airports. It also enables ships to travel safely without risk of collisions. Radar can detect objects as small as an insect or as large as a mountain.

Weather observation and forecasting are helped by radar, which can detect the approach of storms or hurricanes. Scientists use radar to find out about the atmosphere and other planets. It is essential for space travel, allowing controllers on the ground to track craft before they reach their orbits. Radar has many military uses, as it can warn of approaching missiles or attacking planes and ships. On the ground, police use radar to catch speeding motorists.

How radar works

If you shout in a mountain valley, you hear an echo as the sound waves are reflected from a nearby cliff. Radar works in a similar way. It detects things by bouncing radio waves off them. The transmitter of the radar equipment beams radio waves into the sky. When they hit something, such as an aircraft, some are reflected back towards the radar dish or aerial. This passes them on to the receiver, which turns them into an electrical signal. The equipment works out the distance of the aircraft from the time taken for the waves to return. A signal takes 1/500th of a second to return from an aircraft 300 km (190 miles) away. The signal appears as a bright spot or blip on a display screen. This gives the operator information about the position of the aircraft. The radar aerial rotates so that it can cover all the sky.

Most large aircraft have on-board radar to warn them of other planes in the vicinity. It can also warn the pilot about storms ahead, since storm clouds show up on the screen. The radar equipment on board ships beams radio waves out across the surface of the water so that they will be reflected back from any ships or hazards near by.

Scientists are trying to make cheap, small radar units that can be used as aids for the blind or to warn car drivers of hazards ahead.

Flashback

In 1935 the British government asked a scientist, Robert Watson-Watt, to invent a 'death ray' to attack enemy aircraft. He said he could not, but instead he invented the first radar system for detecting enemy planes.

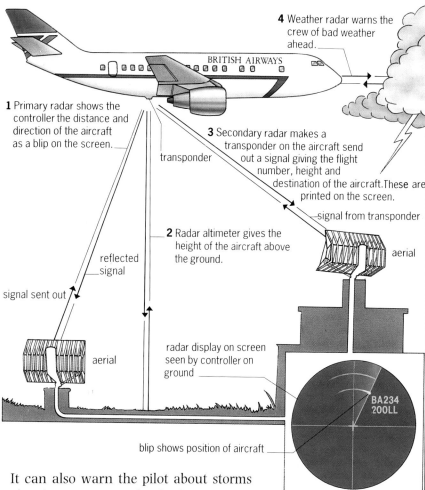

▲ Four types of radar are being used by this aircraft and the controller on the ground. The information on the screen is from the secondary radar. It tells the controller that this is flight number BA234, its destination is London (LL) and it is flying at 20,000 feet (200).

Calculators

Calculators are machines that add, subtract, multiply and divide. Many can do more complicated calculations. They are used by students, engineers, accountants and anyone who works with figures.

At one time, calculators were mechanical. They used rods, levers and gearwheels to do their working out. Nowadays, most calculators are electronic. They work using tiny electric currents. They are very fast, and they do not make mistakes unless you feed the numbers in wrongly.

How they calculate

You can subtract 3 from 5 by finding the number you must *add* to 3 to make 5. And you can multiply 5 by 3 by *adding* three fives: 5 + 5 + 5. Electronic calculators work in a similar way. They use very fast addition to add, subtract, multiply and divide.

When you press the buttons on a calculator, you use ordinary numbers like 0, 1, 2, 3, 4, 5, 6, 7, 8 and 9. The first thing the calculator does is to change these into binary.

▶ Looking inside a pocket calculator. The logic circuits are underneath the middle layer. The bottom layer holds the battery.

▶ Storing numbers in binary. Each number is represented by a sequence of 0s and 1s.

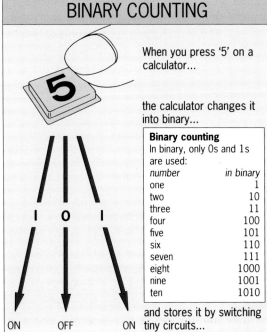

BINARY COUNTING

When you press '5' on a calculator...

the calculator changes it into binary...

Binary counting
In binary, only 0s and 1s are used:

number	in binary
one	1
two	10
three	11
four	100
five	101
six	110
seven	111
eight	1000
nine	1001
ten	1010

and stores it by switching tiny circuits...

ON   OFF   ON

Binary is a method of counting which uses just two numbers, 0 and 1. These are easy for the calculator to handle, because its tiny circuits can be OFF for 0 and ON for 1. The calculator does all its working out in binary. Then it changes back to ordinary numbers to display the answer.

Electronic calculators

Electronic calculators have a tiny logic circuit inside them. These are the 'brains' of the calculator. They run on electricity. They do all the working out, then display the answer on a small screen. Most electronic calculators have a memory so that you can store your answer and use it again later. Some electronic calculators are rather like computers. You can give them a list of instructions, called a **program**. The program can make the calculator do a series of complicated calculations one after another.

There are two types of electronic calculator. **Desktop calculators** can usually be plugged into the mains. Some have a printer in them so that you can print out the answers on paper. **Portable calculators** are small enough to fit in your pocket. Some are as small as a credit card, and some are so small that you can wear them on your wrist in a digital watch. Portable calculators need very little power. Their tiny batteries last for years. Some do not need batteries at all. They use solar cells which turn light into electricity.

Electronic calculators in use

A simple calculator is useful for checking bills or helping with your mathematics at school. But for some jobs, more advanced calculators are needed. Often, these are specially designed to suit particular jobs. Scientists and engineers use specially designed calculators; so do accountants. The electronic checkouts in supermarkets are calculators specially designed to work out customers' bills. On some, you do not even have to press the buttons. The price labels are read automatically by a laser light beam.

Flashback

The **abacus** (counting frame) was probably the first calculating machine. It is an oblong frame holding wires on which beads are strung. On a Chinese abacus there are seven beads on each wire: two above and five below a crossbar. Each bead above the crossbar represents five. Each bead below represents one. The first wire on the right is the units column; the next wire is the tens column, and so on. The thirteenth wire represents trillions. To add you move the beads to the bar, and to subtract you move them away. When you run out of beads, you swap five ones for a five, or two fives for a one in the next column. The abacus is still used in China and Japan today. Skilled operators can work it almost as fast as an electronic calculator.

Mechanical calculators

Before electronic calculators were invented, shops, offices and laboratories used mechanical calculators. These had gearwheels inside them to do the calculating. Mechanical calculators were too large and heavy to be really portable. They were difficult to use and very slow. With some, if you wanted to multiply by five, you had to turn a handle five times. Later versions had electric motors to speed things up, but they were still not as fast as electronic calculators.

The first mechanical calculator was built by Wilhelm Schickard in Germany in 1623. A more famous one was made by the Frenchman Blaise Pascal in 1642. The first calculator to go on sale was the 'Arithmometer' invented by Charles Thomas from France in 1820. By the late 19th century many types of mechanical calculator were being produced.

▲ On an abacus calculating is done by sliding beads along wires. It was probably the first calculating machine but, in skilled hands, can be almost as fast as an electronic calculator.

The first electronic calculator was made in 1963 by the British Bell Punch Company. It was as large as a cash register.

Four years later, Texas Instruments produced a smaller version. This soon led to the cheap, portable electronic calculators we use today.

Information technology

What is Number One in the charts? Who won last night's match? What is the weather forecast? Here are two ways of finding out:

1. Ring one of the telephone information services.
2. If you have TV fitted with teletext, press the buttons on the controller and watch the screen.

▼ A television set fitted with teletext can give up-to-the-minute news of the money markets.

Videotext
Any information service which puts data on your television screen. There are two types:

Teletext services send out data alongside the normal TV signals.

Viewdata services link your TV to the telephone system by computer. You can receive information and send it back again. Home shopping by television is an example of a viewdata service.

Facts like this can be stored in a computer and sent to you by telephone, radio or TV. All these things handle information. They are examples of information technology or IT.

Information technology was probably used when this book was bought. In a large bookshop, the tills are connected to a computer. The computer stores a huge amount of data about titles, authors, prices, books in stock and money taken during the day. This store of information is called a database. A computer can search its database very quickly. It can find the details needed, sort them, display them, and even pass them on to another computer by telephone.

Facts by phone

When computers speak to each other over the telephone, they do not use sounds, as people do. They send data in the form of electrical signals. With a fax machine, pictures and even hand-writing can be sent by phone. At the other end of the line, a second fax machine can receive the signals and print out a copy of what was sent.

The latest telephone lines are not made of metal. They are very narrow strands of glass called optical fibres or 'light pipes'. Signals travel along them as pulses of laser light. They waste less power than the old wires, and they can carry much more data.

Paying with plastic

With information technology, people do not need cash when they buy things. They can use plastic cards instead.

Credit cards have a magnetic strip on them rather like a piece of cassette tape. In a shop, a computer can read the details recorded on this strip. It can check its database to see if the card is out of date or stolen or whether the customer has enough money in their bank account. If all is well, the customer can take the goods.

INFORMATION TECHNOLOGY

The computer sends data about the bill to the credit card company. The customer pays the company later in the month.

Debit cards are rather like credit cards, except that the shop's computer arranges for the bill to be paid straight from your bank account. You pay now and not later.

Prepaid cards are paid for before you use them. Phone cards are like this. You buy a card and use it instead of money when you make a call from a phone box. Each time the card is used, a tiny computer in the phone box changes a number on the magnetic strip. When all the time paid for is used up, the card is no longer valid.

Smart cards are plastic cards with a tiny electronic chip built into them. They can store as much information as a small computer. They can be used as credit cards, debit cards or prepaid cards. They can even store medical details in case of emergency. It all depends on what a computer has told the card to do.

Laser cards now being developed can hold even more information. Their data is read by a laser beam. All the articles in this book could be stored on a single card.

Home shopping with IT

Some people do not like shopping. They would rather stay at home and order goods from a catalogue.

Now, with information technology, they can shop in front of the television. A special computer links the TV to the

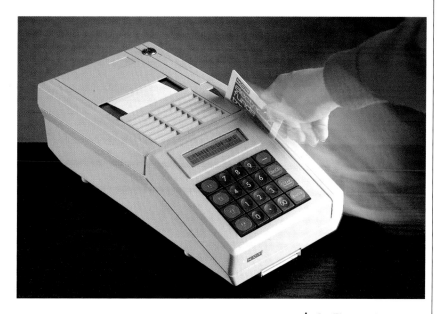

▲ An Electronic Payment Terminal System reads the information held in the magnetic strip on a credit card.

telephone. Lists of goods appear on the screen, just like teletext data. Customers use the computer to select the goods they want. They pay for them by typing in their credit card number.

Working with IT

Many people already use information technology at work. Banks use computers to store account details. Companies use them to work out bills and wages. Travel agents and airlines use computers linked by telephones to make bookings. And at airports, aircraft flights are directed by computers. With information technology, more and more people can work at home instead of going to an office. They can use a computer which is linked to their office by phone. This is called teleworking.

Keeping secrets

Banks, schools, companies and health centres all use information technology to store data about people. Not everyone is happy about this. They worry that computers may pass on information that should be kept private. Many countries have laws to cover the way data is stored and used. In Britain, the law is called the Data Protection Act. It stops organizations passing on information without permission. And it gives people the right to see some of the data being kept about them.

Bar code
Light and dark lines printed on books and other goods. The lines are a code which stands for a number. A computer can use the number to look up the price.

Laser scanner
A machine which moves a tiny beam of laser light over a bar code. A detector picks up the light and dark reflections. It changes them into electrical signals which a computer can understand.

ATM
(Automatic Teller Machine) A machine at a check-out which reads the data on a credit card when the card is passed through it.

EFTPOS
(Electronic Funds Transfer at Point Of Sale) Any system which lets people pay for goods by using a computer at the check-out. The computer arranges for the payment to be made straight from their bank account.

◀ A laser bar code reader in use at a supermarket check-out.

Computers

▶ **Many different devices can be connected to a microcomputer. Some are used for inputting data (green arrows), some for outputting data (red arrows), and some for both.**

Computers are made of several sizes:

Microcomputers (micros) are the small desk-top computers you find in schools and houses. Sometimes they are known as personal computers or PCs. Most can do just one job at a time.

Minicomputers (minis) are larger and more powerful than micros. Some have extra keyboards and screens so that several people can use the computer at once. Each extra keyboard and screen is called a terminal.

Mainframes are the largest computers of all. They have lots of terminals and can do many jobs at once. Most big companies have a mainframe to help them run their business. The computer fills several cabinets and is installed in its own special room.

▶ **Chips inside a computer. The big chips in the middle are the processors.**

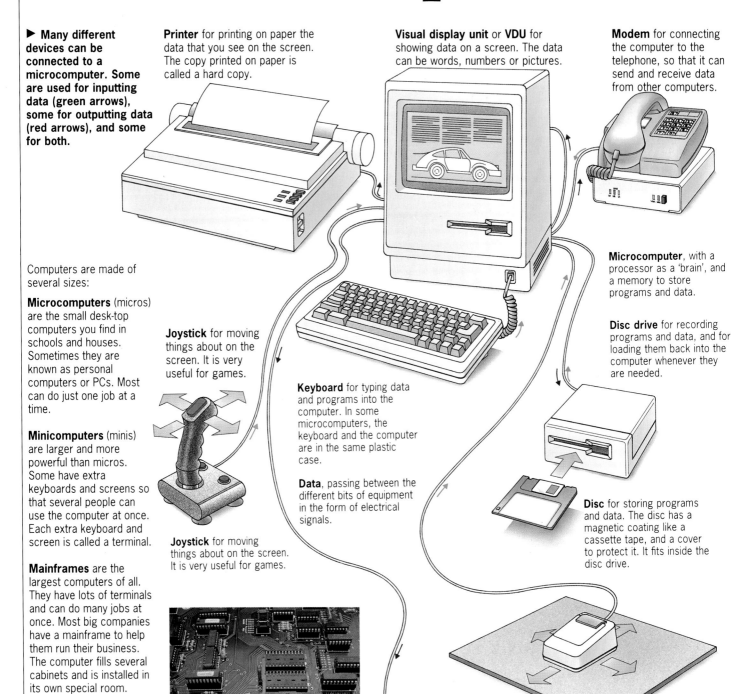

Printer for printing on paper the data that you see on the screen. The copy printed on paper is called a hard copy.

Visual display unit or **VDU** for showing data on a screen. The data can be words, numbers or pictures.

Modem for connecting the computer to the telephone, so that it can send and receive data from other computers.

Joystick for moving things about on the screen. It is very useful for games.

Keyboard for typing data and programs into the computer. In some microcomputers, the keyboard and the computer are in the same plastic case.

Data, passing between the different bits of equipment in the form of electrical signals.

Joystick for moving things about on the screen. It is very useful for games.

Microcomputer, with a processor as a 'brain', and a memory to store programs and data.

Disc drive for recording programs and data, and for loading them back into the computer whenever they are needed.

Disc for storing programs and data. The disc has a magnetic coating like a cassette tape, and a cover to protect it. It fits inside the disc drive.

Mouse for selecting things on the screen and moving them about. It can be rolled in any direction on the desk.

Interface for changing signals into a different form. This interface is letting the computer control a battery-powered model car.

The development of computers

Electronic calculating machines were first made during World War II. The first working computer was built at Manchester University, England, in 1948. From this the first commercial computer, the Ferranti Mark One, was developed in 1951 and filled a whole room. The first computers were large, because their memory circuits used ordinary wires and thousands of valves. Valves looked and glowed like light bulbs. Computers used them to store data by switching them on or off in a certain order.

Computers started to get smaller when transistors replaced valves. Transistors were no bigger than a pea and used much less power than valves. When scientists discovered how to pack thousands of transistors on a single chip, computers could be made even smaller.

Inside a computer

There are thousands of tiny switches inside a computer. They are too small to see and you cannot work them with your fingers. They are electronic switches called transistors and they are turned on or off by tiny electric pulses called signals. The switches are packed together on chips and the computer uses them to store data. It uses a special code for this. For example, it stores the letter N by switching eight tiny switches like this:

OFF ON OFF OFF ON ON ON OFF

The part of the computer that stores data is called the memory. It stores programs as well. The size of a computer's memory is measured in bytes. One byte is just enough memory to hold a single letter or figure. A computer with a memory of 64 kilobytes (64 K) can hold about 64,000 letters or figures.

Data is moved in or out of the memory by the central processing unit (the processor). This is the 'brain' of the computer. It works through the programs, makes decisions and does all the calculations.

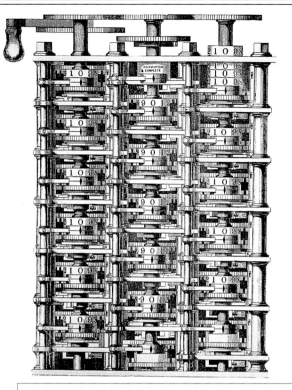

◄ The first calculating machine, invented by Charles Babbage, a British inventor, who is sometimes called the father of the computer. In 1823, he started to build a calculating machine that could do sums and print out the results. The machine used gearwheels to do the calculating. It was so complicated that it was never finished.

HOW COMPUTERS WORK

Imagine you are using a computer to put the names of your friends in alphabetical order. You put the names into the computer by typing them in on the keyboard; you input the data. The computer then organizes and sorts the names; it processes the data. Finally, the computer sends the sorted names to a screen so that you can see the result; the computer outputs the data. Every computer uses these three stages: input–process–output.

Before the computer can sort names, it needs a list of instructions to tell it what to do. This list of instructions is called a program. The computer works through it, line by line, until the job is done. The program can be typed in on the keyboard, but you have to type it in a special language which the computer can understand.

There are many different computer languages. They include BASIC, COMAL, FORTRAN and COBOL. Most home computers use BASIC.

When a computer is switched off, any information you put in is lost. To get round this problem, the computer is connected to a disc drive or a cassette recorder. With this, you can record or save any program or data in the computer. And you can put it back or load it into the computer later, whenever it is needed.

1. Store names in memory
2. Start with letter A
3. Find names beginning with this letter in memory
4. Put names beginning with this letter in order
5. Print names beginning with this letter
6. Stop if letter is Z
7. Take next letter in alphabet
8. Go back to step 3

▲ A flow chart for a program to put a list of names into alphabetical order. Start at the top and follow the arrows.

Robots

▲ R2D2 from Star Wars.

Sniffer robots
Robots that can smell are used to detect gaps in the windows, doors and seams of new cars. These could cause water leaks. The car is filled with air and helium gas, then the robots sniff the seals to detect any gas leaking out.

▶ Robot arms welding on an automated assembly line making Chrysler cars in St Louis, Missouri, USA.

In TV and films, robots are mechanical people with computers as brains. Some are massive, metallic and menacing. They carry laser guns and try to conquer the earth. Others are cute and cuddly like R2D2 of *Star Wars*. Some look so human, you can't tell the difference. These are known as androids.

Industrial robots

There are plenty of real robots on Earth, but they don't look much like people. Usually they are just big arms, fixed to the floor and controlled by a computer. These are industrial robots, used in factories for welding, screwing, cutting, drilling, paint-spraying and packing.

Robots are very good at doing repetitive jobs, where the same thing has to be done over and over again. They never get tired or bored. They make no mistakes. And they can work in conditions which are too difficult or dangerous for people. They can work for hours in heat, cold, dust, damp and choking spray. They can work in the highly radioactive parts of nuclear reactors without falling sick. Robots are also used by the army to blow up mines to make a place safe for people to walk around.

How a robot works

Imagine a factory where a robot arm is packing bottles in a box. The arm can move up, down, and sideways. It has a gripper at the end for picking up the bottles. The gripper can twist and turn, and move in and out.

A computer controls the movements of the arm. It follows a set of instructions called a program. For a job like bottle-lifting, the computer needs signals from the gripper to tell it when to stop squeezing. Otherwise there might be a lot of cracked bottles! Special sensors detect the pressure of the gripper against the bottle and send electrical signals to the computer, just like a human hand sends nerve signals to the brain. The gripper may also need signals

from a video camera to help find the bottles. The computer has a memory to store any programs or signals it is given. Robots can also be controlled from some distance away using radio waves or laser beams to send them instruction signals.

Writing programs for robot arms is very complicated. Sometimes it is easier to give the arm a few basic instructions, then teach it how to move. To do this, someone holds the arm and guides it through all the movements it has to make. The computer stores the details in its memory. Later, the arm can repeat the movements by itself.

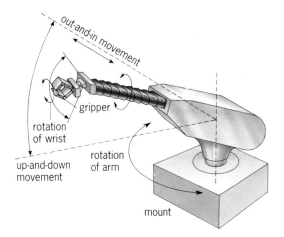

▼ A simple one-arm industrial robot. The gripper can move up and down or, if it is rotated around the wrist, move from side to side. The whole arm, too, is able to rotate on its mount (move sideways) and move up and down.

Robots on the move

Not all robots stay in one place. Robot vehicles have landed on the Moon to explore the surface. And robot spacecraft have photographed distant planets. Some aircraft are like flying robots. They can be controlled by computer, with the pilot taking over for take-offs and landings.

Robot vehicles can be used to move goods round factories. They follow a strip painted on the floor and have special bumpers to stop them when they touch anything.

How robots move

There are three main ways of moving robot limbs. Electric arms are driven by electric motors. Hydraulic arms are driven by oil under pressure, like the shovel on a digger. Pneumatic arms are driven by compressed air. Some robots can recognize where they are in a place, and they can tell how quickly an object is approaching them. This helps them move around without bumping into things.

Seeing and hearing robots

The robots of science fiction films seem human because they can see, hear, smell, touch and even talk like humans. They have sensors, which are structures that can detect the stimulus (the light, sound or smell), and change the sensation into electrical signals that can be understood by a computer. Robot ears, for example, use a microphone that converts sounds into electrical signals. Speaking robots use electronic systems similar to those used in tape recorders to reproduce the sound of a human voice.

Robot sensors can detect things humans cannot, such as toxic chemicals, body chemicals in the blood stream, and magnetic fields. They can touch things without actually making contact, using lasers, X-rays and other ways of sensing things.

Programming a robot
To make a robot arm pick up an object and put it down somewhere else, its computer would need to give it the following orders:
1 Lower arm
2 Close gripper
3 Raise arm
4 Rotate waist
5 Lower arm
6 Open gripper

▲ *Attila*, a robot insect developed in America, can walk across rough ground by itself, and avoid small obstacles using its own logic. It is 30 cm long, and has 23 motors and 150 sensors, including a video camera. Insect-like robots could be used to perform tasks inside complex machines.

Lasers

Laser power
Low power: 1 millionth of a watt. High power: 1,000 million watts. (100 watts runs an ordinary electric light bulb.)

Smallest laser
The size of a grain of salt.

Largest laser
The size of a large building.

Measuring to the Moon
Laser beams bounced off a special mirror, left on the Moon by astronauts in 1969, measure the exact distance to the Moon.

Albert Einstein first suggested in 1917 that lasers would work, but it was not until 1960 that Theodore Maiman made the first working laser.

▶ Inside a helium-neon gas laser.

DANGER
Never look directly down a laser beam. It could damage your eyes and even make you blind.

LASER stands for Light Amplification by Stimulated Emission of Radiation.

Lasers were first called *optical masers*. The invention of lasers was a direct development of the maser (Microwave Amplification by Stimulated Emission of Radiation).

The idea to produce a laser was put forward in 1958 by maser inventor Charles Townes.

Lasers produce a narrow beam of bright light. The beam is much narrower than that which an ordinary lamp can produce, and it is different in other ways as well. First, the light has a single wavelength: it is just one, pure colour. Second, the light waves move exactly in time with each other. Scientists say that laser light is coherent (organized). Ordinary light is not coherent. It is a mixture of colours whose waves are sent out in disorganized bursts.

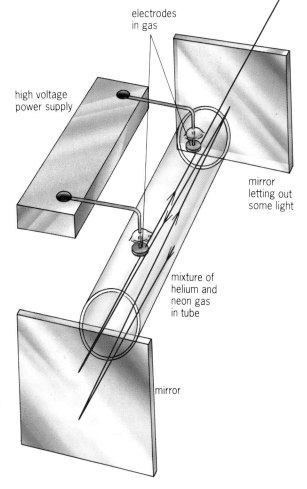

How lasers work

Some lasers have a crystal in them. Others have a tube containing gas or liquid. An electrical discharge or flash of bright light gives extra energy to the atoms in the laser material. The atoms lose this energy by giving out light. The effect builds up as follows. A few atoms lose energy and send out light waves. These trigger other atoms into losing energy and sending out waves . . . and so on. The waves are all identical because they are produced in exactly the same way in every atom. Mirrors at each end of the laser send the light to and fro so that more and more atoms are triggered into sending out light. One of the mirrors is only partly reflecting. Light escapes through it either as a steady beam or as a sudden pulse, depending on the type of laser. The colour of the light depends on the laser material. Some lasers produce invisible infra-red or ultraviolet radiation.

Using lasers

Some supermarket tills use lasers to 'read' the bar codes (the pattern of black and white lines) on the things you buy. The laser beam is reflected from the bars in pulses which are changed into electrical signals. Compact disc players use the same idea. They pick up laser light reflected in pulses from the surface of the disc. Pulses of laser light can also be used to carry telephone signals long distances through optical fibres.

Doctors use lasers to burn away birthmarks and some cancer cells. If the retina (the part of the eye that we see with) comes loose they use a laser to stick it back in place. Military lasers guide missiles to their targets and factories use powerful lasers to cut through metal, glass and even cloth for clothing. Also the coherent light from lasers is used for making holograms.

Holograms

When you take a photograph of an object you just get a flat picture, but if you made a hologram of the same object you would get a 3D picture. This means that the picture stored in the hologram has depth, so you can see round the sides of the object as though it were solid and really there.

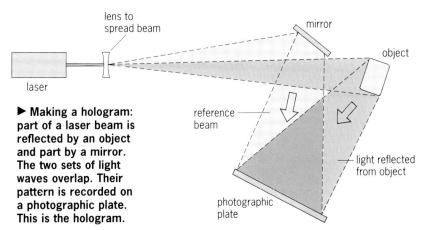

▶ Making a hologram: part of a laser beam is reflected by an object and part by a mirror. The two sets of light waves overlap. Their pattern is recorded on a photographic plate. This is the hologram.

Making holograms

You need the special, single-colour light from a laser to make a hologram. The laser beam is split into two halves. One half shines on the object which reflects the laser light. This light mixes with the other half of the beam (called the reference beam), making patterns called interference fringes. The hologram is a photograph of these patterns. It does not look anything like the original object; in fact you cannot see anything on the hologram plate.

Viewing holograms

To see the picture stored in the hologram you shine a beam of laser light on it: the same kind of laser light that was used to make it. Then, looking through the hologram, you will see the object as though it were really there, although it may have only one colour, the colour of the laser light. You can see some holograms without using a laser, just by ordinary light.

Using holograms

As well as making beautiful pictures, holograms are used for testing in factories. They can show up small differences in things which should be exactly the same. Holography has been used on the Space Shuttle to show which of its protecting tiles have become loose. Some holograms work with reflected light. The holograms used on credit cards are like this. They give a 3D image which is easy to see if you hold the card under a reading lamp. Holograms are put on credit cards because it makes them extremely difficult to forge.

Dennis Gabor first suggested that holograms could be made, in 1948, but it was not until the laser was invented in 1960 that holography really started. The laser produced exactly the right kind of light to make good holograms.

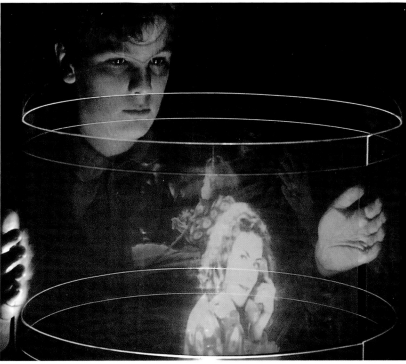

▲ When a circular hologram is illuminated by laser light, it forms a 3D image which can be seen from all sides.

Electronics

▶ In an electronically-controlled simulator like this, pilots can practise their flying skills without ever leaving the ground.

Most of us use electronic equipment at some time. Radios, TVs, disc players and video recorders all use electronics. So do computers, digital watches and calculators. Electronic circuits change and control the flow of tiny electric currents. These tiny, changing currents are called signals. They can make loudspeakers give out sounds and TV screens show words or pictures. They can control machines or carry information. Aircraft navigate by electronics, and automatic cameras work electronically. Heart pacemakers and hearing aids are also electronic.

Electronic circuits

There are four main jobs which electronic circuits can do:

Switching In some circuits, a tiny electrical signal is used to switch another circuit on or off. The automatic doors in shops work like this. When someone comes close, an infra-red detector sends a signal to an electronic circuit. This switches on an electric motor which opens the door.

Amplification This means making weak electrical signals stronger. The signals coming into a radio aerial are very weak. In the radio, an amplifier boosts the signals so that they are strong enough to make the radio give out a sound.

Rectification There are two types of electric current. Current from a battery is direct current (d.c.) it always flows the same way. Current from the mains is alternating current (a.c.): it flows backwards, forwards, backwards, forwards . . . and so on. Rectifier circuits are electronic turnstiles. They let the current through only in one direction. This changes a.c. into d.c.: it rectifies the current. Portable radios run on d.c., but with a rectifier fitted, they can be run from the a.c. mains.

Oscillation The signals picked up by a radio aerial are a.c. The current oscillates (wobbles) backwards and forwards. To handle some of these signals, radios need to produce their own a.c. But the problem is that radio batteries cannot supply it because they are d.c. An oscillator solves the problem. It is an electronic circuit which makes d.c. 'wobble' back and forth so that it becomes a.c.

Integrated circuits

Nowadays, electronic circuits with thousands of parts can be formed on a single chip of silicon no bigger than your finger-nail. The chip is treated with chemicals so that different parts behave like resistors, capacitors, diodes, transistors or connecting wire. The result is an integrated circuit or i.c. It can be the timer in a digital watch, the amplifier in a radio or the 'brain' in a calculator. Microprocessors are the most advanced chips of all. They are used in computers. They can run car engines and drive trains. They can fly aircraft and control artificial limbs.

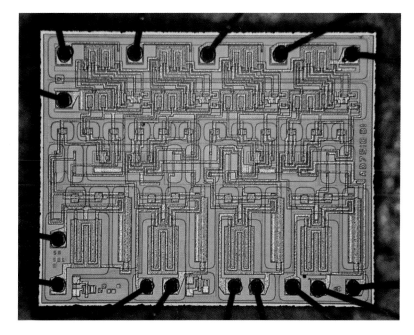

▲ Circuits on a silicon chip, magnified 140 times. The chip is inside a package like the one shown below. Tiny wires connect the chip to the pins ('legs').

integrated circuit

ELECTRONIC COMPONENTS

If you look inside an old radio you will see lots of electronic components. These are the bits and pieces which connect together to make up the electronic circuits. Usually, the components are mounted on a special circuit board. This has copper strips on it to make the connections. The components are fixed to the strips by solder. Here are some of the components you might see:

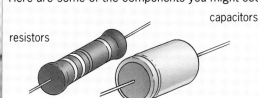

resistors, capacitors

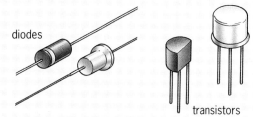

diodes, transistors

Resistors These reduce the flow of current. They make sure that all the other components get the right amount of current to work properly.

Capacitors Some capacitors smooth out the flow of current. Others pass on signals. They are especially useful for this because they do not treat all signals alike. Using capacitors, a radio can be tuned to select signals from one station rather than another.

Diodes These are electronic turn-stiles. They let current through in only one direction. They are used for rectification.

Transistors These are electronic taps. In a transistor, a tiny current can control the flow of a much larger current. It can make the flow stronger or weaker, or can turn it off altogether. Transistors are used for switching, amplifying and oscillating.

Most diodes and transistors are made from a material called silicon. Silicon is a semiconductor. It is not a good conductor of electricity, but it is not an insulator either. To make a diode or transistor, a tiny crystal or chip of silicon is treated with special chemicals. This changes the way the crystal conducts electricity.

Recording systems

Thomas Alva Edison built the first 'phonograph' in 1878 in the USA. The recordings were made on metal cylinders and there were no electronics to boost the sound. The stylus was connected to a large horn.

Records were first used for recording in 1888.

The BBC first used magnetic tape recording in 1932 when they broadcast King George V's Christmas speech to the empire.

You will soon be able to have your photographs put on a Photo-CD quite cheaply, then go home and look at them on your TV screen.

▶ **In a recording studio, 64 separate tracks or more may be recorded alongside each other on the same tape. The tracks can be mixed in any proportion to produce the final sound.**

Most families have at least one recording system in their home. It may be a hi-fi system with records or compact discs, or it may be a video recorder. Families often record their holidays and special occasions with a video camera, making their own video programmes which they can play back through the television screen.

Recording is a way of keeping pictures or sounds (or computer data) so that you can see or hear them later. Cassettes, compact discs (CDs) and vinyl records are all methods of storing recordings. Special machines such as record players and video recorders are needed to play them back. Some machines are called 'recorders' even though they record and play back.

Sound recording

Sounds are produced by air molecules moving rapidly backwards and forwards (vibrating), creating changes in air pressure. To record sound, these changes have to be converted into changing electrical currents. This is done by a microphone. Microphones are found in radio, TV and recording studios, in telephones and in those cassette recorders which allow you to record your own voice.

All microphones have a diaphragm. This is a thin disc of plastic or metal which vibrates when the moving air hits it. A common type of microphone is the condenser microphone. The sound waves strike the metal diaphragm, making it vibrate. The plate has an electric charge. As

INFORMATION TECHNOLOGY

the charge vibrates, it affects the charge on another plate nearby, making current surge in and out of the connecting wires. These surges are the signals.

Putting sounds on tape

Tape recorders can record and play back speech, music, or data from a computer. The recorder contains a reel or cassette of magnetic tape. The tape is wound on to two spools so it can be wound forward or backward. When the player is working, a rotating spindle called a capstan pulls the tape past a record/playback head.

To record your voice, a microphone changes the sounds into electrical signals which are boosted and sent to the head. The tape contains millions of tiny magnetic particles. As the tape passes over the recording head, the head magnetizes the particles strongly or weakly depending on the electric signals it receives. In this way, the signals are recorded as a varying pattern of magnetism in a track along the tape. Professional sound recording tape may have up to 64 different sound tracks on the same tape.

Getting the sound back

To play the recording back, the tape again moves past the head. But this time, the magnetic pattern in the tape generates electrical signals in the head. The signals are boosted by electronic circuits and changed into sound by a loudspeaker. In a loudspeaker a small coil carries the electric current, which reacts with a powerful magnet to make the coil move. The coil is fixed to a diaphragm. As the coil is moved by the current, so the diaphragm moves and creates sound waves in the air.

Stereo cassette players have two loudspeakers. They can record two separate tracks on the same tape. Stereo recordings are more lifelike because the performers seem to be in different positions in front of you.

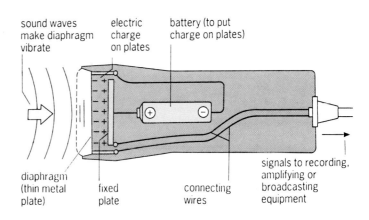

▲ Condenser microphone. When sound waves make the diaphragm vibrate, charge movements send signals along the connecting wires.

ELECTRONIC SYSTEMS

Modern recording systems all use electronics both to record sounds and pictures and to play them back. The recording machine turns the sound and pictures into tiny changing electric currents called signals. When you play back a record, tape or disc, these signals are changed back into sound and pictures again.

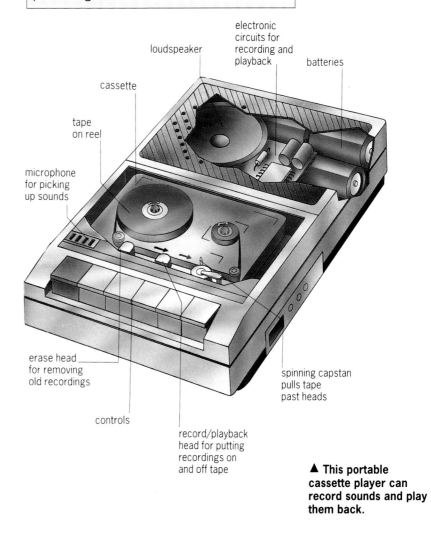

▲ This portable cassette player can record sounds and play them back.

INFORMATION TECHNOLOGY

New kinds of discs and tapes

Laser discs or video discs 2 hours of high quality video, recorded using analogue signals.
CD-I (Compact Disc Interactive) 72 minutes of video, still pictures, sound, text, graphics, animations and computer data.
DVI (Digital Video Interactive) 1 hour of high quality video.
CDTV A 12 cm video disc produced by Commodore.
CD-ROM Video discs that can store lots of data – 20,000 pages of text or 10 copies of a 20-volume encyclopaedia.
Photo-CD 12 cm CD for storing photographs. Produces very high quality pictures.
Recording CD A CD on which you can record.
DAT (Digital Audio Tape) Miniature tapes smaller than a credit card. Player like a very small video recorder. Uses digital signals. CD sound quality. Holds 4 hours of music. Easy to record on or copy.
DCC (Digital Compact Cassette) The player can record and play back both digital sound and ordinary analogue audio cassettes. Plays for only 90 minutes.
Minidisc a 9 cm compact disc that can be wiped, so you can record many times on it. Records 80 minutes of sound, but not quite CD quality. However, unlike CDs, minidiscs are not upset by jolts and bumps, and the players will fit into a pocket. You could use one while jogging.

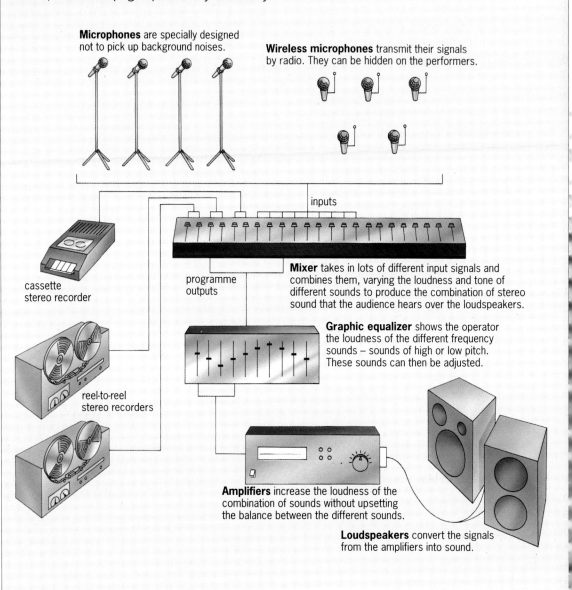

ELECTRONIC SYSTEMS

Live music today requires the use of various electronic devices to make sure the audience can hear every part of the performance. This involves using a lot of different machines. Different parts of the sounds – the soloists, the backup group and any artificially synthesized sounds are all recorded separately. These sounds are then adjusted for loudness and quality, and mixed together to produce the final recording. The diagram shows the main machines used.

Microphones are specially designed not to pick up background noises.

Wireless microphones transmit their signals by radio. They can be hidden on the performers.

cassette stereo recorder

reel-to-reel stereo recorders

inputs

programme outputs

Mixer takes in lots of different input signals and combines them, varying the loudness and tone of different sounds to produce the combination of stereo sound that the audience hears over the loudspeakers.

Graphic equalizer shows the operator the loudness of the different frequency sounds – sounds of high or low pitch. These sounds can then be adjusted.

Amplifiers increase the loudness of the combination of sounds without upsetting the balance between the different sounds.

Loudspeakers convert the signals from the amplifiers into sound.

Recording pictures

Video recording is a way of recording pictures. Most video recorders also record sound at the same time. Like tape recorders, they store information electronically. The signals are stored on magnetic tape, which moves past a recording playback head in the player. Camcorders are video cameras with a small video recorder built into them.

New systems

Recordings like those on records and video tape recorders record the signals as a continuously varying pattern. They are called analogue recordings. But technology has found a better way of storing information. The signals stand for numbers (digits) in a kind of code. The digital signals are changed back to analogue signals before

INFORMATION TECHNOLOGY

being used to make sounds or pictures. Compact discs (CDs) use digital signals. The quality of digital recording is very high because the signals can be copied and transferred from one piece of electronic equipment to another without becoming altered.

Televisions use different 'standards' in different parts of the world. Analogue video tapes recorded on one standard cannot be played on a TV from another system. But digital discs can be played back through any TV anywhere.

Compact discs (CDs)

On a compact disc (CD), the signals are recorded as millions of tiny pits on the surface of the disc. When these are scanned by a laser beam, the reflected pulses of light can be turned into electrical signals and used to make the sound. Video discs use the same principle for recording pictures and sound.

The recording surface of a compact disc is covered in hard plastic. Because it is scanned by a laser beam that does not actually touch the disc, compact discs do not get worn and scratched easily like the old vinyl records, nor is there a hiss if dust gets on the disc.

Interactive discs

Video discs have one big advantage over magnetic tapes. They do not have to be searched from start to finish. The laser beam can scan across the whole disc very rapidly. Video discs can store sound, still and moving pictures, graphics, cartoon animations and text, all on a disc the size of an ordinary compact disc. It can feed sound to a hi-fi and pictures to a TV or a computer screen. It can also be controlled by a computer. You can write a computer program so that the pictures and text will come up on the screen in any order you want. You can make your own lectures, travel shows and so on from the material on the disc. It is easy to search for other pieces of information using the computer or a hand-held control.

Video games

Video games give you action on the screen which you can control yourself, including space invader attacks, Grand Prix races and sports contests. The games are stored on tape, disc or computer microchip in the games machine. The most advanced video games are not really games at all. They are used to train fighter pilots and astronauts.

A 30 cm analogue laser disc holding 2 hours of video costs about £5 to make. A 2-hour VHS video tape costs £1.50 to make. A 12 cm CD holding 70 minutes of video costs just 35p to make.

All European video recorders store and play back 50 separate pictures each second.

Synthesizers are digital machines which can create new sounds and change existing ones. They often have keyboards and you can compose music on them. Then the synthesizer can play it back to you. If you like it, you can save it on a computer disc and print it out as sheet music.

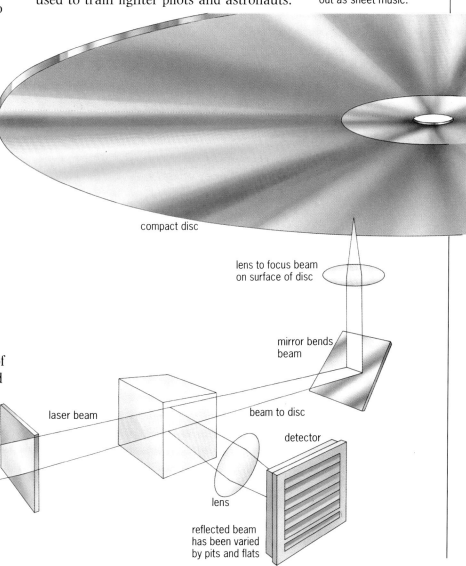

Index

Alphabetization is word-by-word. Numbers in *italic* type refer to captions and illustrations.

A

abacus 141
accumulator 76
acid 11
Acrilan 35
acrylic *34*, 35
advanced gas-cooled reactor (AGR) 71
aerial 132, *132*, *133*, 135
aerofoil 87
aerogenerator 65
aerotrain 96
air 12
Airbus *89*
aircraft *82*, 88-89, *89*
   controls *86*, 87
   forces on 87
aircraft carrier *110-111*
aircraft simulator *150*
airship 90, 91, *91*
Alcock, Sir John 86, 88
Aldebaran 53, 55
Aldrin, Edwin 'Buzz' 40
Alfred the Great, King 30
Algol 53
alkali 11
alloy 32, 40-41
Alpha Centauri 52, 53
alpha-particle 15, 17
Altair 53
aluminium 32, 77
amp 73
amplifier *132*, *133*, 150, 155
amplitude modulation (AM) 132, *132*
android 146
Andromeda constellation 55
Andromeda Galaxy 58, *58*
Antares 52
antimatter 7
Apollo missions 39, 44, *47*
argon 12
Ariane rocket *42*
Arithmometer 141
Armstrong, Neil 40, 41, *41*, 44
Arnould, Michel 90
artificial intelligence 147
asbestos 11
asteroid 47
astronaut *27*, 36, *38*, 39, 40-41, *41*, 44
astronomer 53, 61
atom 7, 8-9 *8*, *9*, 10, *11*, 148
aurora 51
Austin Mini *100*
autobahn *121*
autogiro 93
Automatic Teller Machine (ATM) 143
autopilot 89
axle 28, *28*

B

Bain, Alexander 31, 137
balloon 90, *90-91*
bar code 143, 148
bar code reader *142-143*
barometer 12
barque *108*
bat flight 24, 86
battery 63, *72*, 72, 73, 76, *76*
   car 76, *76*, *101*
Becquerel, Henri 17
Bell, Alexander Graham 136, 137, *137*
Bell, Reverend Patrick 106
belt drive 29, *29*
Benz, Carl *100*
beta-particle 15, 17
Betelgeuse 54, 55
bicycle 107, *107*
big bang 60
binary counting 140, *140*
binoculars 20, 21, *21*
Biopol 35
biplane 89
bird flight 86, 87
black hole 56, 60
Blériot, Louis 86, 88
Boeing 747 94
boiling point 13
Brahe, Tycho 53
brakes, car *101*
Branson, Richard 90
brass 32, 77
Braun, Wernher von 43
bridge 112, 118-119, *118*, *119*
Brindley, James 116
broadcasting *132*, 133
bronze 32, 33
Brown, Arthur Whitten 86, 88
Brunel, Isambard Kingdom 110, 112
Brunel, Sir Marc Isambard 112
bulldozer *105*, *120*
byte 145

C

Cadillac 100, *100*
calcium 11
calculator 140-141, *140*
calendering 35
Callisto 46, 48, *48*
caloric 25
camcorder 154
camera 129, *129*
   television *134*, 135
Cameron, Don 90
camshaft 103
canal 116-117
cancer 17, *17*, 41
candle *19*, 30
capacitor *151*
car 100-101, *100*, *101*
car assembly line *146*
carbohydrate 14
carbon 7, 8, *8*, 10, 11, 14
carbon cycle *14*
carbon dioxide 12, 14
carburettor 103
cassette recorder *153*
Castor and Pollux 53, 55
Cayley, Sir George 86, 88
Celsius (centigrade) scale 13
Centaur 55
central heating 82, *82*
centre of gravity 27, *27*
Ceres 47
chain drive 29, *29*
chain reaction 70
chalk 11
Challenger locomotive 97
chandelier *19*
Channel Tunnel 122, *123*
charcoal 14
Charon 46, 49
chemical formula 11
chemical symbols 10
Chernobyl 71
chip, electronic *144*, 145, 151, *151*
chlorine 7
chromium 32, 33
chronometer 31
circuit 72, *72*, *73*, 75
clay 11
clepsydra 30
clipper (ship) 108-9, *109*
clock 30, *31*
CN tower 122, 123
coal 7, 14
cobalt 77
Cockerell, Sir Christopher 92
cog (ship) *108*
colour 22, *22*, 23
colour printing 23
combine harvester 106, *106*
comet 57
compact disc 152, 154, 155
compact disc player 148
compound 10
compression 26
compressor 78, *78*
computer 89, 142-3, 144-5, *144*
Concorde 36, 85, 89, *89*
condenser 78, *78*, 95
conduction 25, *25*
conductor, electricity 72
constellation 55, *55*
container ship *110-111*, 116
container traffic 104, 115
convection 25, *25*
cooker 80, *80*
cooking 81, *81*
copper 11, 32, 33, 77
Corinth canal 116
cosmic ray 17
cosmonaut (astronaut) 40
Crab Nebula 56
crane 124, *124*
crater 57
credit card 142-143, 149
credit card reader *143*
cruise liner *110-111*
Crux (Southern Cross) 55, *55*
crystal 11, *11*, 14, 40-41
Curie, Marie 17
current, electric 72, 73

D

Dacron 35
Daguerre, Louis 129, 131
Daimler, Gottlieb 102
dam 63, 68, 69, *69*
Darby, Abraham *119*
data 145
Data Protection Act 143
Davey, Christopher 90
Davy, Sir Humphry 97
De Havilland Comet 88, 94
debit card 143
decibel (dB) 24, *24*
Deimos 46
derrick 124
detector *133*
Dewar, James 12
dhow *108*, 109
diamond 7, 11, 14, *14*
diesel oil 14
diode *151*
Dione 49
dirigible 90, 91, *91*
distributor 103
dock 114, *114*
Dog Star (Sirius) 52
Dorigny, Hélène 90
double glazing 83, *83*
double star 53
driveshaft *101*
Duralumin 32
dye 23
dynamo 77

E

Eagle (star) 55
Earth 36, 44-47, *45*, *47*, 59, *59*, 60
　gravity 27, *27*
　heat store 67
　magnetic poles 77
earthquake 122
Eastman, George 129
echo 24
eclipse 51, *51*
Edison, Thomas Alva 152
EFTPOS computer system 143
Eiffel tower 122
Einstein, Albert 148
electric current 15, 151
electric shock 74
electrical insulation 83
electricity 62, 68, 72-73
　solar powered 66
electricity supply 74-75
electrolysis *33*

electrolyte 76
electromagnet 77
electromagnetic wave 15, 16, *16*
electron 8, *8*, 15, 72, *72*, 73, 134-135, *134*, *135*
electron beam 15
electronics 150-151
element 8, 10, *10*, 32
elephant communication 24
Enceladus 49
energy 9, 15, 25, 62-64, 69, 70-71, 75
　electrical 63, 73
　geothermal 64
　kinetic 62, 63
　nuclear 52
　potential 62, 63
　Sun's 50
　water 68
energy sources 62-63, *62*, *63*
energy use 64, *64*
engine, diesel 96, *96-97*, *101*, 103, 105
　electric 96
　internal combustion *101*, 103
　jet 94, *94*
　motor cycle 102
　railway 96, 97, *97*
　ship 112
　steam 95, *95*
　turbine 94
engine capacity 103
ethanol 11
Europa 46, 48, *48*
evaporation 78
extrusion 35
eye 20

F

facsimile (fax) 137, 142
fan belt 103
ferry 85, 110, *111*, 115
fertiliser 11
film, camera 129, 131
fire, discovery 81
fission, nuclear 9, *9*, 70
flight 86-87
flow chart 145
fluorescent tube i
Flyer I 86
Focke, Heinrich 93
force 26-27, *26*, 29
Ford, Henry 100
foundation 122
freeway *120-121*
freezing point 7, 13

frequency 16, *16*, 132-133
　mains 73
　sound 24
frequency modulation (FM) 133
friction 25, 26, *26*, 62
fuel, engine 103
　fossil 14, 62, 63, 64
　nuclear 52
　rocket 42, 43
fuse 75
fuse-box 75
fusion, nuclear 50, 52, 71

G

Gabor, Dennis 149
Gagarin, Yuri 40, 41
galaxy 58, *58*, 60, *60*
Galileo Galilei 30, 46
galleon *108*
gamma ray 15, 17, *17*, 70
Ganymede 46, 48, *48*
gas 7, *7*, 9, 10, 11
　bottled 63
　natural 14
gas mantle 19
gases in air 12
gear, bicycle 107
gearbox, car *101*
gearwheel 29, *29*
Geiger counter 15, 33
Gemini (Twins) 53, 55, *55*
generator *70*, *71*, 73, *73*, 74, *75*, 77
geyser 67
glider 86
gold 8, 11, 32, 33, 77
Gossamer Condor 86
Graf Zeppelin airship 91
graphic equaliser *155*
graphite 7, 11, 14
gravity 26-27, 52, 60
　Earth's 36
　Moon's 44
　Sun's 45
gravure 126, *127*
Great Bear (Ursa Major) 55, *55*
Great Britain (ship) 110, 112, *112*
Great Eastern (ship) 112
Great Nebula 56
Great Pyramid 122
Great Red Spot 48
Great Western (ship) 112
grid, electric 74
Gutenburg, Johann 127

H

Halley, Edmund 57
Halley's Comet 57, *57*
Hampson, William 12
Handley Page HP42 *82*
hang glider *87*
Harrison, John 31
harrow 106
heat 7, *7*, 16, 25, 62, 83
Heinkel HE178 88
helicopter 93, *93*
helium 12, 90, 91
Helix Nebula *53*
Herschel, William 49
Hertz, Heinrich 17
High Speed Train (HST) *97*
Hindenburg airship 91
hinge 28
hob 80
hobby-horse 107, *107*
Holland, John 113
hologram 148, 149, *149*
Hooke, Robert 31
Hoover Dam *68*
hot spring (geyser) 67
hovercraft 85, 92, *92*
human body 13, 14
Huygens, Christiaan 30
Hydra constellation 55
hydrocarbon 14
hydroelectricity 68, 69
hydrogen 8-13, 52, 90, 91

I

ice *7*, 13
ignition coil 103
Illinois Waterway 117
information technology 142-3
infrasound 24
insect flight 86, 87
insulation 83, *83*
insulator, electricity 72
integrated circuit 151, *151*
interface *144*
Io 46, 48, *48*
ion 8
iron 8, 11, 32, 33, 77
isotope 9, 10
Issigonis, Alec *100*

J

jack, car 28, 29
Jessop, William 116
jet aircraft, fastest 87, 88
jet engine 89, 94, *94*

Jodrell Bank 61, *61*
joule 62
joystick, computer *144*
juggernaut 105
junk (ship) *108*, 109
Jupiter 39, 45, *45*, 46, 48, *48*, 49

K

Kepler, Johan 53
Kiel Canal 117
kinetic theory 25
kitchen range *80*
krypton 12

L

La France airship 91
Lake Mead *68*
lamp 19, *19*, 30
Land, Edwin 129
laser 148, *148*, 149, 154
laser card 143
laser scanner 143
Lavoisier, Antoine Laurent 12, 66
lead 8, 32, 77
lens 20-21, *20*
Leonov, Alexei 41
letterpress 126, *127*
lever 28, *28*
light 16, 18, 22-23 *22*, 62
  speed 15, 18, 59
light bulb *19*, 75
light waves *18*
light year 39, 52
lighting 19
lightning *72*
lightning conductor 123
Lilienthal, Otto 88
lime (chemical) 11
Linde, Karl von 12
Lindstrand, Per 90
liquid 7, *7*, 10, 11
lithography 126, *127*
lock (waterway) *114*, 116, *117*
locomotive 96-97, *96-97*
lorry (truck) 104-105, *105*
loudspeaker 132, *133*, 153, *153*, *155*
Luna spacecraft 44
Lunar Module 39
Lunar Rover 39, 41
Lunokhod robot 39
Lyre (star) 55

M

M49 galaxy *58*
M4 star cluster *52*
McAdam, John 121
machine 28-29
machinery, farm 106, *106*
Magellanic Clouds *58*
maglev train *96*
magnesium 32
magnet 73, 77, *77*
Maiman, Theodore 148
mangle *79*
Marconi, Guglielmo 133
Mariner spacecraft *46*
Mars 39, 45, *45*, 46, 47, *47*, 49
maser 31, 148
mass 27
matter 7-8
melting point 7
Mensa star *55*
mercury (metal) 32
Mercury (planet) 45, *45*, 46, *46*
metal 32-33
meteor 52, 57
Meteor Crater 57, *57*
meteorite 44, 57, *57*
Michaux, Ernest 102
Michaux, Pierre 102
microphone 132, *132*, 152-3, *153*, *155*
microprocessor 151
microscope 20, 21, *21*
microwave *16*, 62
microwave oven *80*
Milky Way Galaxy 58, 60
Mimas *49*
mineral 11
mixer, sound *155*
modem *144*
modulator *132*
molecule 7, *7*, 9, *9*
  heated 25
  long-chain 34
  water *13*
monoplane 89
Montgolfier, Jacques 90, 91
Montgolfier, Joseph 90, 91
Moon 36, 44
  eclipse 51, *51*
  exploration 39
  measuring to 148
  phases *44*
Moon buggy *44*
moons, orbiting planets *46*
motor cycle 102, *102*
motorway 120, 121
moulding 35

N

nebula 56
neon 12
Neptune 39, 45, *45*, 46, 48, 49
neutron 8, *8*, 9, *9*, 15
neutron star 56
Newcomen, Thomas 95
Newton, Sir Isaac 26
nickel 32, 77
Niépce, Joseph Nicéphore 129, 131
Nott, Julian 90
nuclear energy 62
nuclear fission 9, *9*, 70
nuclear fusion 50, 52, 71
nuclear power 70-71
nuclear radiation 9
nuclear reactor 9
nuclear waste 71
nucleus, atom 8, *8*, 9, *9*
nylon 34, *34*, 35

O

observatory 61
odometer 101
Ohain, Hans von 94
oil 10, 14, 34
oil platform *122*, 123
oil tanker 110, *110-111*
optical fibre 142, 148
orbit *45*
ore 32, 33
Orion 54, *54*, 55, *55*, 56
oscillator *132*, 151
oxygen 8-13, 42, 43
oxygen mask *12*

P

paint *23*
Panama Canal *116*, 117
paper 128, *128*
paraffin 14
Parkes, Alexander 34
Parkesine 34
Pascal, Blaise 141
Pavo 5 galaxies *60*
penny-farthing 107, *107*
Perseus *55*
petrol 14
Phobos *46*
phonograph 152
photocopying 127
photography 130-131, *131*
photon 18
photosynthesis 12, 62
piston 95, *95*
pivot *28*
planet 36, 45, *45*, 46-49, *46*, *47*, *48*, *49*
plastics 34, *34*, 35, *35*
platinum 32
Pleiades 54, *54*, 55
plough 106, *106*
Pluto 39, 45, *45*, 46, 49
plutonium 15, 71
poles, magnetic 77
polonium 17
polymer 34, *34*
port 115, *115*
power 62, 68-71
  electric 74, *74*
  geothermal 67
  nuclear 70-71
  water 68-69
power station 65, 73, 74
  geothermal 67, *67*
  hydroelectric 68, *68*
  nuclear 70, *70*, 71
pressure 12
pressurized water reactor (PWR) *70*, 71
printer, computer *144*
printing 125-127
  colour *23*, *126*
  halftone 126, *126*
prism 18, *18*, 21, *21*, 22
program, computer 145, 146-7
propeller 89, 110
proton 8-9, 15
Proxima Centauri 52
pulley 28, *28*, 29, *29*
pulsar 53, 56
PVC 34, *34*

Q

quantum mechanics 8
quartz crystal 31
quasar 56

R

radar 139, *139*
radiation 15-17, 41
  background 71
  electromagnetic 16
  gamma 70
  heat 25, *25*
  laser 148
  light 18, 62
  nuclear 9, *9*, 15

sound 17
radiator, car *101*, 103
radio 132-133, *133*
radio wave *16*, 21, 132-3, *132*, 135
radioactivity 9, 15-17, 71
radium 17
radon 12, 17
railway 98-99
   Great Western 112
railway station 99
railway track 98, *98*
rainbow 22
reactor 70, 71
recording 152-155
recording studio *152*
rectifier 150-151
red giant 53
refrigerator 78, *78*
resistor *151*
Rhea 49
ring main 75, *75*
rings, orbiting planets 46, *46*
ripple 18, *18*, 22
road 120-121, *120*
robot 29, 146-147, *146*
rocket 36, *36*, 42-43, *42*, *43*
The Rocket (locomotive) 97
Romanenko, Yuri 41
rust 33
Röntgen, Wilhelm 17

S

sailing 108-109, *108*, *109*
St Lawrence Seaway 117
salt 7, 11, *11*
sand-glass 30, *30*
satellite 36, 37, *37*, 42, *42*, 46, 138, *139*
Saturn 39, 45, *45*, 46, *46*, 49, *49*
Saturn 5 rocket 39, 42, 43
scales *26*
Schickard, Wilhelm 141
schooner *108*
sensor 147
Seven Sisters 54, *54*, 55
ship 110-112, *110-112*
ship, sailing 108-9 *108*, *109*
shooting star 52, 57
signal, railway 99
silicon chip 151, *151*
silver 11, 32, 33, 77
simulator, aircraft *150*
Sirius 52, 53, 55
skull, human *15*
smart card 143
smelting *33*
soda 11

sodium 7
solar cell 37, *38*, 66
Solar Challenger 64
solar flare *50*
solar furnace 66
solar panel 64, *64*, 66, 138
solar power 66
Solar System 45-51, 60
solid 7, 10, 11
sonic boom 24
sound 17, 24, 62
sound broadcasting *132*
sound insulation 83
sound recording 152-153
Southern Cross (Crux) 55
space exploration 36-40, 46
space probe 36, 39, 46
Space Shuttle 36, *36*, 37, 39, 149
space station *38*, 39, 40, *50*
spectrum 18, *18*, *22*, 23
speedometer 101
Sputnik 1 *138*
star 52-54, *52*, *53*, *54*, 60
starch 14
steam 7, 25
steam-engine 95, *95*, 97
steamship 109, 112
steel 32, 33, 77
steering, car *101*
Stephenson, George 97
Stephenson, Robert 97
stereo sound 153
STOL aircraft 89
submarine 113, *113*
Suez Canal 117
sugar 10, 14
sulphur 11
sulphuric acid 11
Sun 16, *16*, 44, 45, 50-53, *50-51*, 59, *59*
   eclipse 51, *51*
   gravity 27
   power source 66
sundial 30, *30*
sunspot 51
super-gravity 60
supernova 17, 53, 56, 60
supertanker 85
suspension, car *101*
Sutcliffe, Frank Meadow *130*
switch, electronic 150
symbols, electricians' *73*
synthesiser 155

T

tachometer 101
Talbot, William Henry Fox 129, 131

talc 11
tanker *108*
tape recorder 152, 153, *153*
Taurus (The Bull) 54, 55, 56
telefax 137
telephone 136-137, *136*
telescope 20-21, *20*, 39, 61
teletext 142, *142*
television 134-135, *134*
teleworking 143
Telford, Thomas 116
Telstar 138
temperature 7, 13, 25
temple 122
tension 26
Tereshkova, Valentina 41
Terylene 35
Tethys 49
thermodynamics 25
thermos flask 12, 83
Thomas, Charles 141
threshing machine 95
tidal power 68, 69
time 30, 31
tin 11, 32, 33, 77
Titan 46, 49
Tompion, Thomas 31
torch 76
Townes, Charles 148
tractor *105*, 106, *106*
trailer *104*
train 96-97, *96-97*, *98-99*
transformer 74, *75*
transistor 133, 145, *151*
transmission line 74, *74*
transmitter 132
transport 84, *84*, 85, *85*
Trevithick, Richard 95
Triton 46, 49
truck 104, 105, *105*
tuner *133*
tungsten 32
turbine 67, 68, *69*, 70, 71, 73, *73*
   steam 110, 112
turbine engine 94
type 125, *125*
tyre, bicycle 107

U

ultra high frequency (UHF) 135
ultrasound 24
ultraviolet rays 50
Universe 59, *59*, 60
uranium 8-9, 15, 17, 32, 70
Uranus 36, 39, 45, *45*, 46, 48, 49
Ursa Major 55, *55*

V

V-2 rocket 43, *43*
vacuum cleaner 79, *79*
vacuum flask 12, 83
valve 145
Venus 36, 39, 45, *45*, 46, 47, 49
very high frequency (VHF) 133
vibration 25
   sound 24
video recorder 154, 155
videotext 142
Viking spacecraft 39, *47*
Volkswagen Beetle *100*
volt 73
voltages, typical 72
Vostok spacecraft 41
Voyager spacecraft 39, 46, 48, *48*, 49
VTOL aircraft 89

W

washing machine 79, *79*
watch 30, 31, *31*
water 7, *7*, 9, 10, 13, *13*
water power 69
water wheel 69, *69*
Watson-Watt, Robert 139
Watt, James 95
wavelength 16, *16*, 18, 21, 22, 133
weather data 90
weather forecasting 138, 139
weight 12, 26, 27
weightlessness *27*, 40, 41
wheel 28, *28*
   train *98*
Wheeler, John 60
Whittle, Sir Frank 86, 94
wind power 65
windmill 64, 65
wing 86, *86*, 87, 89
Wright brothers 86, 88, *88*
Wright Flyer 88

X

X-ray 15-16, *15*, *17*, 50, 60, 62
X-ray camera 129
xenon 12

Z

zinc 32, 33